# Location Awareness in the Age of Google Maps

*Location Awareness in the Age of Google Maps* explores the mundane act of navigating cities in the age of digital mapping infrastructures.

Noone follows the frictions routing through Google Maps' categorising and classifying of spatial information. Complicating the assumption that digital maps distort a sense of direction, Noone argues that Google Maps' location awareness does more than just organise and orient a representation of space—it also organises and orients imaginaries of publicness, self-sufficiency, legibility, and error. At the same time, *Location Awareness in the Age of Google Maps* helps to animate the ordinary ways people are challenging and refusing Google Maps' vision of the world. Drawing on an arts-based field study spanning the streets of London, New York, London, Toronto, and Amsterdam, Noone's encounters of "asking for directions" open up lines of inquiry and spatial scores that cut through Google's universal mapping project.

*Location Awareness in the Age of Google Maps* will be essential reading for information studies and media studies scholars and students with an interest in embodied information practices, critical information studies, and critical data studies. The book will also appeal to an urban studies audience engaged in work on the digital city and the datafication of urban environments.

**Rebecca Noone** is an artist and a Lecturer (Assistant Professor) of Information Studies in the School of Humanities at the University of Glasgow.

# Location Awareness in the Age of Google Maps

Rebecca Noone

Routledge
Taylor & Francis Group

LONDON AND NEW YORK

First published 2024
by Routledge
4 Park Square, Milton Park, Abingdon, Oxon OX14 4RN

and by Routledge
605 Third Avenue, New York, NY 10158

*Routledge is an imprint of the Taylor & Francis Group, an informa business*

*British Library Cataloguing-in-Publication Data*
A catalogue record for this book is available from the British Library

ISBN: 978-1-032-17049-7 (hbk)
ISBN: 978-1-032-17050-3 (pbk)
ISBN: 978-1-003-25156-9 (ebk)

DOI: 10.4324/9781003251569

The Open Access version of chapter 1 was funded by University of Glasgow.

# Contents

# Acknowledgements

Thank you to everyone who helped me bend with the elasticity of time while writing this book—the long, deliberate stretches and the sudden snaps of fervent energy. The process seemed slow and steady as well as all-of-a-sudden and spurting. Thank you to everyone who held me together, and helped me stay in one piece.

Thank you to Heidi Lowther and Heeranshi Sharma and everyone at Taylor & Francis for seeing this project through, from the initial discussions about this book through to its completion. Thank you to the Social Sciences and Humanities Research Council of Canada for generously funding my postdoctoral and doctoral studies, where I developed this project.

I am especially grateful to Karen Dewart McEwen who helped me gather my fragmented thoughts and structure them into a book. And I am so thankful for the support of Jenna Hartel who encouraged this book from the very beginning. I am also deeply appreciative of all those who took the time to read and comment on chapters or listened to early presentations of this work, or through their own brilliant writing, helped me imagine the type of research I wanted to do. Thank you to Dylan Mulvin, Dani Metilli, Emily Maemura, Katie MacKinnon, Aparajita Bhandari, Rianka Singh, Jess Lapp, Arun Jacob, Jamila Ghaddar, Mariam Karim, Vanessa Fleet Lakewood, Sarah Switzer, Madison Trusolino, Camille-Mary Sharp, Elisha Lim, and Sara Bimo.

This project, now a book, has taken on many forms. It was an art project long before it was a research project, and, as such, owes its life to the curators, residency programmes, visual methods enthusiasts, and arts collectives who gave it (and me) space to be weird. Namely, I would like to thank Shannon Linde, Patricia Ritacca, Emily Fitzpatrick, and Renée van der Avoird at Aisle 4 and the Art of the Danforth Festival in Toronto, the people at The Luminary Gallery and Artist Residency in St. Louis, Mark Dunford and the International Visual Methods Conference community, and the people at YTB Gallery and the Roundtable Residency. Perhaps most importantly, I am deeply appreciative of all the people who stopped and gave me directions over the course of performing this project across a number of cities.

I developed this art project into a research programme at the University of Toronto. Thank you to the mentorship of Jenna Hartel, Jas Rault, T.L.

Cowan, Nicole Cohen, Irina Mihalache and Leslie Regan Shade. I am also grateful to everyone at the Digital Research Ethics Collaboratory, the Critical Digital Media Institute at University of Toronto, Scarborough, and the Critical Digital Humanities Initiative. The biggest thank you to Sarah Sharma and the life she brought to the McLuhan Centre (which truly changed how I think). I am also so grateful to Matthew Brower who helped me navigate a pathway between research and art. Thank you.

Thank you to everyone at the Department of Information, University College London which was my homebase for two years as a postdoc. I am especially thankful to the sharp wisdom of Annemaree Lloyd and Alison Hicks who read, commented, and mentored. I am also so grateful for Elizabeth Shepherd for welcoming me to the department and for Charlie Inskip for continuing to cultivate a supportive academic space. Also, the biggest thank you to Adam Crymble, Steve Gray, and Lucy Stagg and everyone at UCL Digital Humanities Centre for your collaborations and guidance.

Thank you to my family who love and care in all the ways they can: Gerard, Joel, Ann, Mary Rose, Malila, Willie Ann, and Jo. I am grateful to friends Jessie, Shannon, Matty and Amy. Thank you for being funny when things seem less funny. Thank you to Tara for helping me check the facts. Thank you to the friends who I shared life and space with–Colin, Stef, Sylvie, Cath, and Marion—who witnessed and soothed. Finally, thank you, Dylan—the person who animates and enlivens thinking and expressing and experiencing the world. Who helped me push this book to the end. Rather than depleated, I am primed and ready to draw some new lines. Thank you.

# 1 The Lost Art of Location Awareness

## Two Starting Points

*Starting Point A:* In 2018, I walked through the cities of London, Amsterdam, New York, and Toronto; and, as I walked, I stopped passers-by to ask for directions. I would ask how to get to the near and the far away: local sites that so often sit patiently in the background, such as libraries, markets, or parks; and those iconic city attractions that proudly stamp their presence over tourist brochures, like the Tate Modern or Heineken Experience.

More specifically, I would ask the people I stopped to draw directions for me, offering them paper and pen. These moments produced a series of hand-drawn vernacular maps; line drawings that translated an abstracted spatial memory to paper. Some drawings are minimal layers of lines, Xs, text, and arrows while others are dense spatial indexes—some plain and direct, while others scrawled and perhaps disorienting. Together, the hand-drawn spatial scores plotted out a route in a *here* to *there* of location awareness.[1]

*Starting Point B:* In 2019, Professor Bradford Parkinson and his collaborators received the esteemed Queen Elizabeth Prize for Engineering to honour their role in developing the now seemingly ubiquitous Global Positioning System satellite navigation network, better known simply as GPS.[2] The Elizabeth Prize, one of the top prizes in engineering, is just one of Parkinson's many awards—the touted "hero of GPS" also boasts a Marconi Prize and the Draper Award, among many others.[3] Parkinson's accolades hardly come as a surprise with GPS now seemingly sutured into the fabric of navigation. Operating on both a global and individual scale, GPS facilitates the tracking of fleet shipments and coordinates deliveries, while it also helps orchestrate everyday wayfinding when enabled through mobile mapping platforms like Google Maps.[4]

But amidst the celebration of a world forever changed by this locative technology, Parkinson was circumspect. Apart from the Queen Elizabeth Prize feting, Parkinson was making headlines for his misgivings about GPS's widespread application. In an interview with Tom Whipple, Science Editor at *The Times*, Parkinson concedes that while he is proud of GPS as a technological

DOI: 10.4324/9781003251569-1

development, the pervasive use of GPS-enabled mobile mapping platforms like Google Maps needed to be approached with caution and not blanket celebration. In the words of Parkinson, "There are downsides; every advance has that. The fact is that people don't know how to read maps anymore."[5] As summarised by Whipple, Parkinson "worries about the lost art of map reading as people turn to GPS-powered maps on their smartphones."[6]

## The Lost Art of Reading the Map

For Parkinson, the emergence of GPS in everyday navigation practices causes a deficit of map reading skill. Now the map simply *knows* and does the navigational work. For Parkinson, the location awareness of GPS opened up a new type of vulnerability: What if the GPS system should be jammed? Or hacked with intentionally misleading information? Or infiltrated for surveillance? Not being able to read a map left an opening for susceptibility to so-called malicious interventions by some "bad actor" who can take everything down in one "bad faith" manoeuvre.[7]

Parkinson's suspicion reflects his own background in militaristic intelligence and defence. As part of the US military, Parkinson helped develop a global navigation system using extra-planetary satellites—this became the foundation for the GPS of today. In 1973, Parkinson became the first Director of NAVSTAR GPS to build out intelligence "on the ground."[8] Like other geolocative technologies Graphic Information Systems (GIS), GPS enmeshed the specificities of precise location tracking in a totalising vision of the world, expressly designed for the purposes of military tactics.[9] As GPS became commercial and consumable, embedded in cars, phones, pet collars, and toys, its knowingness became commonplace. Nonetheless, it carried with it this omniscient precision of locational information charged with a command over what Caren Kaplan terms, the "target subjects."[10]

Today, Google Maps is perhaps the most prominent example of GPS presence in everyday life.[11] It is one of the most used digital mapping platforms with almost 2 billion monthly users and nearly 70% of the digital mapping market share (followed by Waze, which Google acquired in 2013).[12] Google Maps uses GPS's "location-enabled" technology to help locate, route, recommend, and coordinate. GPS literally positions people on Google Maps between one metre and thirty metres of their exact location (though Google's official measurements of accuracy are difficult to track).[13]

GPS is also fundamental to how Google builds and operates its map. In the background, Google uses GPS to help organise the spatial information it accumulates and assembles. For example, Google Maps uses GPS to locate and pin Google Street View images to Google Maps.[14] Google operators and software extract rich spatial details from Street View imaging—like addresses of businesses, one-way streets, or traffic lights and stop signs—and input these details into Google Maps located via GPS coordinates.[15] The Elizabeth Prize recognised Parkinson and his role in GPS precisely for this ubiquity and embeddedness of the locative technology—the same embedded-ness spurring Parkinson's concern.

While the Frankenstein conceit—the inventor penitent for their creation—may seem prosaic, Parkinson is not alone in his anxieties about compromised navigational skills and mobile map dependencies. Parkinson joins a chorus of concerns, including the laments of journalist Michael Harris, who in his 2015 book, *The End of Absence: Reclaiming What We've Lost in a World of Constant Connection*, argues that pervasive GPS (namely mobile maps) means that *we don't know how to be lost anymore*, let alone how to read a map. According to Harris, plugging into GPS's network of spatial intelligence has resulted in what he considers a loss of basic human survival skills of finding one's way.[16] Additionally, Harris bemoans what he sees as the dearth of playful serendipity that animates intuitive, rather than computational, navigation—a worry echoed in Stephen Petrow's 2018 road diary in the *USA Today* titled "I was a GPS Zombie. Here's what happened when I went back to paper maps and serendipity."[17] Petrow recounts his adventures of hitching his way across America without a mobile map with the goal to "relearn" how to "connect with people" in the process—a testament to all the experiences that are supposedly missed when finding one's way with satellite navigation systems.[18] For Parkinson, Harris, and Petrow, GPS and its attendant mobile applications are disconnecting us from our environment rather than helping us move through it.

Another feature of the satnav lament is the suggestion that GPS has fundamentally altered neurological processing, producing a cognitive lack of *awareness* about location. In 2017, Cari Romm, writing for New York Magazine's *The Cut*, warned that "using Google Maps Too Much Really Does Mess with Your Sense of Direction."[19] The previous year Greg Milner for *The Guardian* penned the headline: "Death by GPS: are satnavs changing our brain?" to pathologise cognitive function at the hands of convenient mapping tools.[20] Importantly, the common enemy throughout this handwringing is not simply GPS as a global system, but also mobile maps such as Google Maps, Waze, and Apple Maps, as an individual experience of these systems. As such, not using these mapping platforms, or going rogue on road trips, is something that interrupts the ubiquity of systems, or as Romm suggests, sets a course for retraining one's brain. But this presumes that one can opt out of using digital mapping systems and platforms lock stock, without accounting for what Jean-Christophe Plantin is Google Maps increasingly infrastructural role.[21]

Parkinson's commentary—like that of Romm, Petrow, and Harris—is made newsworthy precisely because it converts the narrative from "GPS as convenient tool of location awareness and navigation" to "GPS as a trap." Terms like "zombie" and "death by GPS" imply a threat to survival at the hands of insufficient navigational skills. But implicit in this criticism is the assumption that navigation is "an innate skill" and, moreover, navigation is a normative cognitive function under threat. The individual is then likely to be duped by bad actors, as Parkinson suggests, or lose connections with their

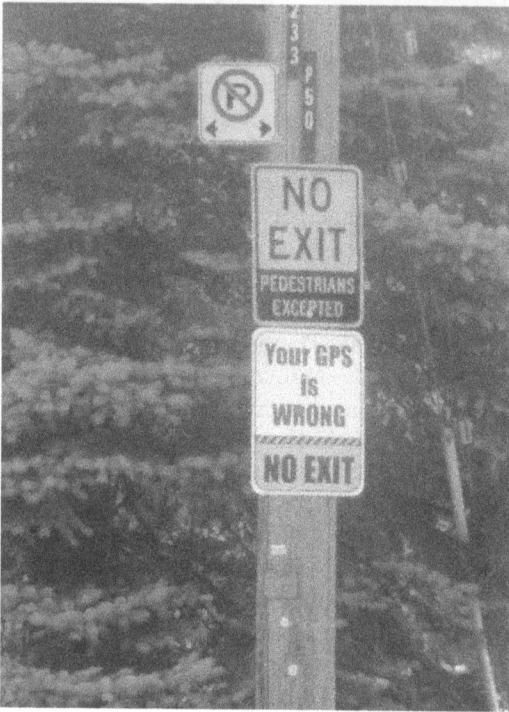

*Figure 1.1* "Your GPS is wrong. No Exit" sign posted by neighbourhood residents in Toronto. Photo by Emily Maemura.

environment, as Harris fears. Seduced by the ease of digital maps and bereft of spatial awareness, we can neither be lost nor found (Figure 1.1).

## A Sense of Direction

Regardless of how one might feel about their sense of direction, how one navigates space is always a question of processing, sorting, and directing a multitude of information with and without digital maps. Jason Farman calls this multifacted practice of being in space a "practice of sensory-inscription"—which he defines as the "proprioceptive sensing" of the immediate environment in tandem with one's embodied and socio-cultural situation.[22] For Farman this type of sensing is a form of "implacement" that is "culturally inscribed and contextually specific."[23] In other words, a sense of direction follows multiple routes. Therefore, the continual coordinating within a sense of direction reflects Annemaree Lloyd's theorisation of information practices oriented by a personal "information landscape."[24] Information landscapes are the interconnected information systems and tacit forms of knowledge, layered with community practices and contextual reasoning that are all part of negotiating a situation or an environment.[25] Proprioceptive sensing unfolds within the complexities of information landscapes.

Situating the information landscape of location awareness in the context of mobile technologies and screen cultures, Nanna Verhoeff argues that the digital map interface and the city are "dialogic encounters between visual, virtual, material and physical domains, and as such operate as space—the time-binding set-ups or dispositifs of performative navigation."[26] Instead of zombies or deskilled navigators, Verhoeff activates the digital map user as engaged in what she calls "performative cartography" in which information on and off screen become an interplay of meaning, braided in the gestures, the interfaces, the movements of navigating the city. Moreover, the digital map works alongside the other features of the mobile phone including messaging, taking photos, tagging locations, augmenting other forms of information. Interfaces and their translations to digital maps reflect Alexander Galloway's understanding of interfaces as *processes* in and of themselves, rather than simply objects.[27] It follows that a sense of direction is a weaving of multiple strands of phenomenal processes, occurring simultaneously and perhaps in paradox. It is an unfolding practice rather than a declarative knack.

Rescaling a sense of direction to the everyday navigations of situated and structured knowledge upholds Doreen Massey's conception of space: that space is *relational* rather than *dimensional*.[28] Massey argues that "if 'space' as a dimension is anything at all, it is the dimension of coexisting actors, the dimension that precisely enables (and requires) their multiplicity."[29] Key to space's dimensionality are the variable relations to power. As such, spatial experiences are co-existently situated and embodied practices and strategies that both absorb and refuse the hegemonic systems of power imbalance.[30] For Fran Tonkiss, this means navigating space while contending with differential experiences of *being seen* in space or feeling safe.[31] These spatial tactics play out empirically in actions like having a personal shortcut, but it can also be taking the *long way* to feel safe on busy streets, or taking a shortcut to feel safe in not being seen. These routings change and adapt based on who circumscribes this path, and when and where they do it.

Understanding space as uneven and differently coded is important for thinking about the production of senses of direction. For example, Katherine McKittrick's Black geographies conceptualise a poetics of space activated in the practised resistance to pervasive racist geographies. She writes that "Black women's lives are underwritten by ongoing and innovative spatial practices that have always occurred, not on the margins, but right in the middle of our historically present landscape."[32] McKittrick illustrates how racism is central to how space is thought of and imagined, while contemporaneously Black geographies are deeply embedded in spatial meaning-making and practices of occupying space.[33] These inform one's sense of direction and movement through space. The presumption of a "lost sense of direction" loots space of its very histories, narratives, and positions.

By reterritorialising a sense of direction as occurring within an information landscape—a landscape produced through the tensions of structure and embodied geographies and the "proprioceptive" sensing of place—it becomes clear that one does not easily relinquish directional skill at the hands of Google Maps. The goal of the book is to bring to the fore a politics often lost in these popular critiques of Google Maps—popular critiques like "Google Maps has ruined our sense of direction" and "no one can read a map anymore!" Rather than navigate the platitudes of vanquished senses of direction or gained conveniences, this book tracks another way through the spaces Google *maps* and the attendant assumptions that location awareness is a stable, unidirectional phenomenon. Moreover, Google Maps is a sticky social object that carries with it the histories of other mapping projects that have been used to delineate boundaries in the name of property, assign valuation and accumulation, and target "from above."[34] Despite all Google's claims to mapping innovation, wrapped up in fantasies of location awareness, so much of Google's drive for location awareness reinforces already entrenched power structures and the relations to place they engender. The book dredges out the functions and fictions of a Google-positioned location awareness—its claims as a public resource, to promises of self-sufficient exploration, to templating representations of space, to indexing value—that organise and orient relations to space on and off the map.

**Orienting Location Awareness**

In 2006, Malcolm McCullough argued that information is increasingly about you and about *where* you are.[35] This information tailoring became increasingly poignant with the emergence of smartphones, shortly following McCullough's declaration. Mobile media carried with it the promise of information access about *anywhere*—one could be anywhere![36] And while this idea of accessing information anywhere might suggest a collapsing of space and a non-specificity of location, Eric Gordon and Adriana De Souza e Silva's 2011 book, *Net Locality: Why Location Matters in a Networked World,* argues that location plays a central role in how information is organised and navigated online.[37] They contend that net locality, or location awareness, in the age of mobile computing is foundational to how "we *navigate* information" and moreover "the way we expect to be *navigated*."[38] But instead, in this age of mobile computing, location remains important. Broadly speaking, location awareness is part of accessing place-based information and also spatialising and locating search queries. This is what Gordon and De Souza e Silva refer to as the "re-territorialisation" of space because "we are where our devices are."[39] Instead of a non-specificity of placeness, smartphones marked what Gerard Goggin and Larissa Hjorth have termed "the locational turn in mobile technology."[40]

While location has come to signal precision of place—a specific site of being—location is also an indefinite concept of relating to space and place; at once informative of, transcended by, or used interchangeably with presence and positionality. As Adriana de Souza e Silva and Jordan Frith argue, location in the context of locative media like digital maps, and beyond, exists in a tension between "fixed geographical coordinates" and "complex multi-faceted identities that expand and shift according to the information ascribed to them."[41] Indeed, Anne Galloway and Matthew Ward, and Minna Tarkka, among others, have all argued that location and location awareness are active and dynamic rather than a static point on a map that folds in time, infrastructural arrangements, and sensory perceptions.[42]

Beyond moments of receiving location-aware information, using mobile phones also engages location-specific processes of sorting and selecting. As Jordan Frith and Didem Özkul argue mobile technologies like smartphones, as well as Walkmans, iPods, and barcodes, can be a means to "negotiate a certain type of control over their spatial experience."[43] Pinning locations from one's holiday abroad, sharing location details for safety, or triangulating location tags on photos to jog a memory, are just some of the ways by which, to adopt Germaine Halegoua's term, we "re-place" space through Google Maps.[44] Here re-placement is not a takeover of spatial sensemaking, but as Halegoua writes, "re-placing is a set of practices that manage the seemingly fragmented and overwhelming conditions that the networked urban subject experiences and routinely acts within, then re-embeds these conditions within meaningful spatial and temporary contexts."[45] Or, to put it another way, location awareness does not evaporate when one uses a mobile map, but is augmented, annotated, or even affirmed. In this way, using mobile maps like Google Maps is not a wholesale surrender of individual awareness but is, what Sarah Barns identifies as selective and iterative engagement with space.[46] Digital mapping brings to the fore the elasticity of location awareness. Indeed, location awareness is expansive as it binds together the fragments of experiences of moving through the world.

We all carry with us personal location awareness, activated in mundane and urgent ways that put us in place and ensure safe passage. In many cases, this location awareness is tied not only to navigating space but also to navigating spatialised relations to hegemonic forms of power, or what Doreen Massy terms power geometries.[47] Power geometries show how power manifests spatially, through who claims a right to space, who is "out of place," and who assumes safety in that space. These power differentials reflect the structures and systems of white, patriarchal heteronormativity. Spaces like cities are sites that Sarah Elwood dubs "divisive socio-spatialities" which she argues become amplified in increasingly mediated spaces premised on white supremacy, settler coloniality, and heteronormativity.[48] These translate to other forms of location awareness, like that described by Rinaldo Walcott in his account of walking through Toronto as a queer Black man and the anticipations of being read as "out of place" by police.[49] It can also be found in Isabel Waidner's *Sterling Karat Gold* where the title character, Sterling, practices a routine scanning of the area around their estate in Camden to clock that which threatens their passage as a queer, non-binary person. And, as Waidner's story imparts, this location awareness can also be knowing where care is, where the refuge is, and where one can find solidarity in resistance.[50]

Location awareness is a relational and responsive means of moving through space. It is affective, embodied and tied to survival. Location awareness is not simply an affordance of the map, it is its fault lines that serve a presumed standard experience of space. The types of location awareness like that of Walcott and Waidner dislodge the worry of losing a sense of direction to Google Maps. Location awareness is also the way to contend with the gated-off fantasies of public good, self-sufficiency, spatial legibility, and accurate representation that Google's brand of location awareness perpetuates.

## Programming Location Awareness

Google Maps' locational prowess and its rise as what technology writer Andrew Hawkins declares is "go-to navigational tool of our time" was not inevitable;[51] but, as Scott McQuire, Rowen Wilken, Mark Graham and Martin Dittus have respectively shown is the result of large-scale investments in the development and acquisition of technologies, processes, and protocols of translating space to data.[52] The product of these ventures is an expansive database of geographic and location-based information, from satellite imagery to street maps, to 360-degree panoramic views of Street View images, real-time traffic updates, and route planning for pedestrian, car, bicycle, and public transportation read through the kaleidoscope of Google Maps' location awareness.

As Scott McQuire observes, "when Google Maps began in 2005, Google was a late entrant to the field."[53] Other forms of early personal digital navigation are in the form of MapQuest (owned by AOL), Yahoo! Maps, and Windows Live Local (the precursor to Bing Maps). These maps came with the novelty of turn-by-turn directions provided as text and symbols beside the graphical map, marking an easy visual reference for each point in pathway decision-making. Additionally, one could print off the directions and bring them along with them—to carry their individualised route while in transit, prototyping what would become normalised with the advent of the mobile digital map.

Under the direction of Bret Taylor, Maps' co-creator, Google Maps was to reorient the map around one's ever-changing position rather than a from a single, static place, tailored to an information search within a specific moment.[54] In 2003, Taylor managed an early Google project known as "Search by Location" that "finds" a location with the entry of a ZIP code and a keyword. While this search system was an early and albeit less functional Google Maps (one that Taylor describes as "practically a useless project"[55]) what was key to Search by Location was the centrality of location paired with search—and imagining of the world as searchable. The idea was that the map should not be static, like the traditional paper map or the printed

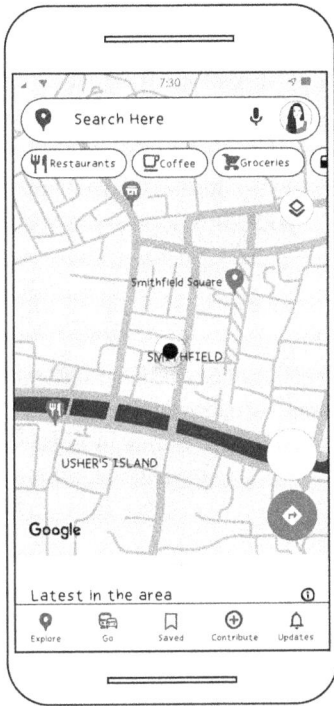

*Figure 1.2* Illustration of Google Maps' interface. Illustration by Colin Medley.

off directions—but dynamic. That map was to be interactive in content and form, addaptive to adding new information, zoomable, rotatable, and endlessly scrollable. This dynamic interaction with the map marks a shift away from the map as page and towards the vector map and its aggregated tiles that make the representation of space responsive and searchable. How one interacted with the map and *saw themselves on the map* mattered. (Figure 1.2).

To achieve this vision, Google acquired several other geomedia startups, along with their developers—including Brian McClendon, Lars and Jens Rasmussen and Mark Grady—to lead the development team of Google Maps.[56] Indeed, many of the features and affordances synonymous with the Google Maps of today were developed "out-of-house" and acquired by Google. Companies such as Where2Technoglies, Keyhole Corp, Zipdash, and SkyBox Imaging became enmeshed into the Google Maps interface.[57] Part of Google's manoeuvring was to make these not only available to subscribers of previous platforms but to the public through the desktop for free.[58] Google's acquisition of companies such as Keyhole seemed to grant access to spatial data and representation, and the location awareness that it afforded.

Keyhole's main product was *Earth Viewer*, a digital mapping database of satellite images. Keyhole's vector organisation of its locational data, portioned out via glimpses through the door (as its company name suggests), was part of how Google Maps came to claim to complete coverage.[59] Skybox was an interface-scrolling technology. Skybox helps Google standardise the experience of using the digital map—what it felt like to smoothly move through the earth. Together, Google could stick images together, enabling them to seamlessly be scrolled through and zoomed in and out of, turning the static digital map into one that is easily moved through. It was responsive and adaptive, enabling a sense of boundless "exploration" with the scroll function.[60] ZipDash uses GPS to track vehicles and analyze speed and traffic conditions. Initially it was developed for highways in Los Angeles, San Diego, Seattle, San Francisco, and Phoenix.[61] ZipDash was a peer-to-peer network for sharing traffic conditions, reflective of the age of other systems like Napster and Kazaa, but tied to "wireless location-specific information. Using cellphones to deliver information and advertising to users in specific locations."[62] Location awareness was enacted through this gesture of pinching and scrolling, zooming in, total views, and ability to know the best route. Keyhole, Skybox, ZipDash, among other spatial computing programmes became part of how Google Maps platform that gave *everyone* access to *everywhere*—thus entrenching its status as a "public" resource as well as helping to standardised expectations for what looking at a digital map looked and felt like, combining the visual language of vectors with a responsive interface and the endless scroll.

Another key part of Google's mapping trajectory was to build a searchable representation of the world. In 2010, Google integrated Google Maps into its search engine meaning that whatever was entered into the Google search engine was visually located with an embedded map at the top of the search results,[63] providing a direct link to the Google Maps application or webpage. According to Joel Kalmanowicz, Project Manager at Google Maps, one in every five searches is location related; and on mobile, almost one in every three searches is location related.[64] Location awareness was a process of information sorting and data organisation, mediated by information about space.

More than just a map, it could be a platform for other services that require location-based information. Soon after its launch in 2005, Google released Google Maps Application Programming Interface (Google Maps API) as a free model for software and platforms to use. Google Maps' API enabled people, most notably businesses, to design maps using Google Maps' template, and personalise them to their respective personal and commercial interests.[65] This (initially) "free" API meant that it was easy to access. It quickly became the go-to mapping software for people and businesses, effectively universalising digital mapping protocols and means by which location and data were connected.[66]

Google Maps' API was part of the democratisation of the geolocation industry which was until then specialised in geographic research and for the military. Gordon and de Souza e Silva write that Google's API meant that "the specialised domain of the GIS programmers became the domain of everyday users."[67] And, indeed, Google Maps' open-source API became the most used mapping API. And while this idea of "the nonexpert" mapping became entangled with the democratising of the map, Jean-Christophe Plantin identifies that it is this seemingly free and easy-to-use API that facilitated the centralisation of Google Maps' infrastructure and effectively the enclosure of its systems.[68] The Google Maps API extends the reach of Google through its base map and its related software. This is part of how the map becomes woven into other platforms, extending the interface of the map to other applications such as car-sharing platforms, delivery services, and real estate sites.[69]

Google frames location awareness beyond simply locating and being located through its map. Location awareness is part of the democratisation narrative of digital maps, with *you* at the centre.[70] Google Maps stakes its claims to a democratic map through claims to access and personalisation—the promise that everyone can use the map anywhere in the world.[71] And even better, one doesn't have to find oneself on the map to begin navigation (*you are here*), the map identifies that location and tailors the map accordingly. Google's My Maps feature, according to Google Maps' About Page, helps users "easily create custom maps with the places that matter to you."[72] The democratic map is made democratic via your position and interests. Moreover, Google Maps offers interaction and participation through its Local Guides program, adding reviews and verifying information *local to you* such as store hours and business closures. The map situates representation of the world as an activation of territory rather than a representation.[73]

### Reorientating Location Awareness

Beyond a technical function, Google's promise of location awareness is a cultural claim encoded with imaginaries about space and mapping. These include the object-based imaginaries about the map's operations—the proficiencies of map production, the accuracy of the image itself, and the experience of reading and applying the maps' directional acumen—as well as the fantasies of world buildings, including the promise of the democratisation of cartography, the ease and fluency of finding one's way, and the prowess to build a complete map of the world. Crucially, imaginaries are more than benign ideas but active modes of organising and directing in the world, baked into sociotechnical systems like Google Maps. Claudia Strauss argues that imaginaries are not necessarily themselves common social practices and ideas but "*make possible* common practices" as well as "effect a shared sense of legitimacy" within those practices.[74] The project of location awareness is a project that assumes a total map is possible but also assumes that one can invest in and claim space through the map.

According to Lucy Suchman, imaginaries are an active part of both designing and using technologies because they are always informing and mediating the relationship between what a technology presents as doing and what it does.[75] It subsumes information and fills a vacuum of available information. For example, Taina Bucher uses "algorithmic imaginaries" as a framework for examining the algorithmic cultures of Facebook—ways of thinking about how algorithms are popularly conceived in the absence of concrete, publicly available, and broadly understandable information on algorithms.[76] Understanding algorithms as a cultural artefact rather than a purely technical one means being attuned to these affective dimensions of algorithms—how they are talked about, promoted, *imagined.* Imaginaries follow the same bearings on the map.

Imaginaries can be restrictive and binding, as much as they can be expansive and elastic. Benedict Anderson's concept of imagined community identifies how imaginaries structure values of the collective that might undermine a sense of collectivity by stratifying who and what belongs to "the community."[77] So while imaginaries might appear to be held in common, they also structure and inform the cleavages of social life, becoming a type of implicit cultural model of the world. In this sense, imaginaries are operatative and affective. To take up Lauren Berlant, imaginaries ensare a politics of publicness, as both the *feeling* of connection as well as the *unfeeling* of normativities.[78] In this sense imaginaries both prevent and produce deliberations towards a just future or "a collectively invested form of life."[79]

Lilly Irani and Ruha Benjamin have respectively critiqued the imaginary of *innovation* for its laminations as an inevitable driving force set on improving life. Irani demonstrates how fantasies of innovation as collective good is a conceit that, in turn, renders necessary the exploitation of a globalised workforce.[80] Also in critique of inflated innovation, Benjamin has shown how chasing the new and the innovative to manage social ills becomes an easy way to overlook the structural racism at the helm of these systems.[81] The result is an uneven distribution of innovation's so-called benefits yet a blanket assertion that innovation is a wholesale good.

Turning to imaginaries of spatial technologies, Lisa Nakamura dispels the fantasy that spatial and immersive technologies enable new modes of awareness and orientation.[82] For example, Nakamura questions Meta's vision that virtual reality headsets are a technology capable of "promoting empathy" such as showcasing the destruction of Hurricane Maria in Puerto Rico to people who did not experience the natural disaster. Nakamura shows that this use of VR further instils distance from the remote "other" and turns social urgency and the geopolitics of aid into spectacle. Nakamura demonstrates how such VR technologies are used to collapse space, and doing so, make new claims on knowing the world. In other words, imaginaries organise and orient who belongs. Imaginaries can structure how one relates to or feels about technology; and imaginaries also conceal the social arrangements supporting technologies and their systems.

Beyond the technological imaginary, imaginaries inform relations to space and thoughts about space. These form what Doreen Massey terms the "geographical imagination" or the implicit and explicit conceptualisations of space. In *World City*, Massey demonstrates how an imaginary of London as a financial centre, where wealth is concentrated, often erases the lived reality of being poor in the city. These imaginaries shape assumptions about what engagement with space looks and feels like. These "implicit geographies" of place are ultimately parts that are "made to stand in for the whole."[83] Geographic imaginaries, like technological imaginaries, organise and orient—they classify and categorise space as well as assume some trajectories are inevitable.

For Sara Ahmed, orientations can be a lens through which to consider the relation between spatial experiences and power since orientation shapes the relationship between space and action. Ahmed takes up orientation as a type of "queering of phenomenology" or a means to think through how bodies are "straightened" and "directed" by constructed norms and pervasive spatial logics but also how resistance to these is an orientation.[84] Ahmed writes that orientations "shape not only how we inhabit space but how we apprehend the world of shared inhabitancies, as well as 'who' or 'what' we direct our attention toward."[85] Combined, orientations are the actions and phenomena of spatial imaginaries, of being in and moving through, while the organisations become the geographic ontologies of space, the contours, the boundaries, and the openings. Location awareness operates within this territory.

Both geographies and orientation get at the question of who and what is at the centre of Google Maps' spatial imaginaries and claims to organise spatial information to make it understandable and accessible.[86] This book asks: what are the orientations and geographies baked into this claim and the relations to space they forestall and foment? And while this book mostly uses the language of Google Maps to speak about digital mapping projects, this is not meant to be at the exclusion of other geo-related Google products like Google Maps and Google Earth, or other geomedia from Apple Maps to OpenStreetMaps. While they have different uses, different functions, and different looks, there is a through-line of presumed global totality. The goal of the book is to challenge the claims of objectivity and universalism of spatial representation while considering the prototyping and spatial conditions that make these claims seem possible. The automation of spatial decision-making is an issue not of getting lost but of losing the organisation of space to a private, consumption and ownership model of big tech. Google Maps positions computational location awareness as the organising principle of the local rather than the local—a site of situated knowledge—as the organising principle of location awareness. These reveal spatial formations that happen out of sight but are practised in the everyday experiences of moving through spaces.

## The Art of Location Awareness

The ideas I explore in this book developed from an arts-based research project about wayfinding inspired by a work by artist Stanley Brouwn titled *this way brouwn.*[87] In 1961, Brouwn walked the streets of Amsterdam, approaching other pedestrians and asking them for directions to nearby sites such as the centrally located Dam Square or City Hall. He offered the helpful stranger a piece of paper and a pen for them to draw their spatial instructions. Brouwn performed the actions of *this way brouwn* over and over—quietly accumulating drawings of directions.

In *this way brouwn,* Brouwn's material is space—the tuning into space and the experience moving through the city's forms and contours. Brouwn's interruptive action has the imaginative possibility of depicting a moment when one locates oneself and communicates that sense of location through drawing and narration. Looking at the markings now, they appear as a series of lines, nodes, and incomprehensible scribbles. Each drawing represented a set of directions, with their own internal orientations and geographies, imagined in a spontaneous exchange. Put together, they are a spatial record of how to get from A to B in a complex, information-rich environment. These rudimentary maps are not for perfect navigation or even for being a comprehensive representation of space. While Brouwn exhibited this work and published a selection of maps as part of a book, these drawings of space do not come together to compose a total map of Amsterdam; but, instead, by often repeating routes, Brouwn's visual transcriptions destabilise the idea of a total space if not the simple futility of defining a single (or optimal) route through space.

Brouwn's art plays with the possibility of art being woven into the commonplace, like asking for directions. Tomas Schmitt calls Brouwn's work a set of "real actions" that reflect ordinary encounters,[88] while curator Claire Lehmann calls them "quiet actions that are not necessarily legible as art to an onlooker."[89] As a piece of performance art, it is relatively *quiet,* demanding little attention; but nevertheless, it is an interruption of the streets' taken for granted flows. As a series of objects—the drawings—they are curious and evocative but relatively unusable as a map or re-performable as a set of directions beyond the moment of exchange. Instead, the drawing leaves one wondering: Where is the starting point? How does one decipher the lines? What does that x represent? Instead, the site-specificity of each encounter serves as a reminder of the mediated experiences of the world in a movement that both abstracts and deeply personalises experiences of the surround.

Over 50 years after Brouwn began performing *this way brouwn,* a project he performed on and off until his death in 2017, I reactivated the performance. I describe this project in the opening lines of the chapter. I start with a similar act of asking for directions in London, Amsterdam, New York, and Toronto. I requested people draw directions for me using the paper and pen I provided. In my version of *this way brouwn,* I collected hundreds of drawings from these moments of spontaneous spatial sense-making and on-the-spot navigation.

Like in *this way brouwn,* these encounters were deliberately subtle situations, meant as a type of disappearing event. In some sense, they echo the work of

Guy Debord and the Situationists International who sought to move through the world with the intention of awareness, engaging with the density and multiplicity of spatial encounters.[90] However, the flaneur is also an unreliable narrator, as Fran Tonkiss argues, "the flaneur in the nineteenth century linked forms of spatial practice with a certain kind of masculine subjectivity."[91] This subjectivity was effectively a masculine entitlement to see everything without seeming out of place for looking. In response, I temper the flaneur with the irreverence of what Robert Filliou terms good-for-nothingness to acknowledge that limitations and absurdity of mapping as a means to disrupt an instrumentalisation of the map.[92] The maps are, for the most part, not transferrable beyond the encounter nor are they even transposable in space. Instead, they make sense in tandem with the experience of asking for directions and watching the abstraction of space unfold in a drawing. Nor, are the maps the only way to get to the destination, nor promise to be the best or the fastest. Often, I was told, this route was the clearest to explain or the easiest to follow. Sometimes, the drawings depicted what I found to be *the wrong* route or details might be confused and disorienting.

At the same time, I was becoming a repository for people's confessions about using digital maps like Google Maps. People would speak to their use of Google Maps followed up by a passing "shame on me, I should know," or sometimes there was apologising for using Google Maps with statements such as, "sorry, I have to use Google Maps, it is easier this way." Or there were moments of "thank goodness for Google" since Google knew the way or "do you not have Google Maps?" as a comment about interrupting their day. In other words, the stuff around the drawings was just as fascinating as the drawings themselves. It was the fragments and idiosyncrasies that both made the exchanges memorable and animated the spaces. These events also destabilised the idea of a complete view of the city even mediated via Google's directions. These directions improvised a reading of space that interpreted and reinterpreted Google Maps or parsed out aspects of Google's directions while devising a different reading of space on top of those directions (Figure 1.3).

The drawings and encounters initiated a line of questioning I build upon throughout this book. However, the research itself is limited to a specific cultural context: four Euro-American cities that are socially and culturally diverse while also deeply implicated in colonialism and its ongoing violence. These cities do not represent a universal wayfinding experience even though they are often cities where Google Maps tests its products (such as Immerisve View in London). Additionally, the perceptions, directions, and gestures that manifest in the city streets cannot be adequately captured in a picture, nor can the meaning of their movements be circumscribed in a text. Google Maps is not available everywhere and the full suite of Google Maps' affordances and products is only available in a small percentage of cities, among which these four cities belong. My focus on these cities is therefore profoundly limited in a way that reflects the bias inherent in Google Maps' global mapping projects: these four cities are centres of global economic power; they

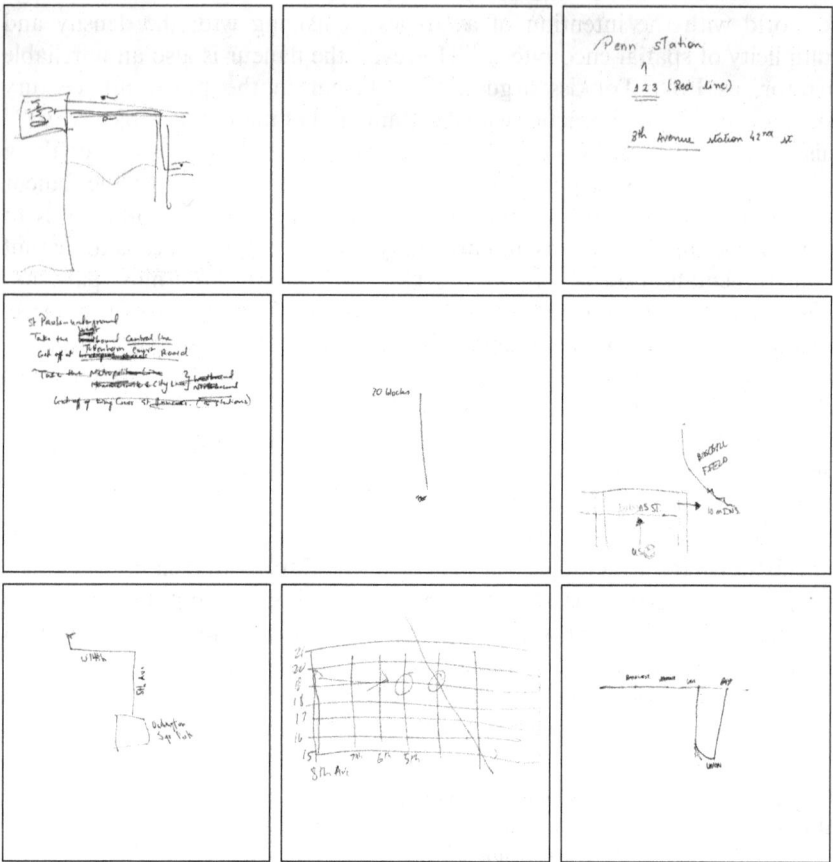

*Figure 1.3* A selection of nine drawn directions collected by the author from Amsterdam, London, New York City, and Toronto.

are hubs of both digital and physical infrastructure; they privilege navigation by English speakers. This book highlights how these kinds of spaces, which have been built to uphold the dominance of whiteness and "the West," have come to serve as models for "the city" within Google Maps' infrastructure. London, New York, Toronto, and Amsterdam are often painted as global or multicultural cities, a discursive framing that optimises difference and flattens the experiences of the people who live there. It also obscures the barriers to access these cities put in place that disproportionally harm racialised, trans, queer, poor, and other marginalised people. The complexities of these experiences and how they are reflected and refracted through wayfinding strategies are not addressed in the primary data collected for this book, but critical race, queer, and feminist theories play a grounding role in the book's intervention in Google's spatial imaginaries. Parsing out the discourses and practices of Google Maps in relation to these moments of asking for directions is a means to plot out the geographies and orientations baked into its location awareness. It calls to mind the productive tensions of location awareness. While the act of asking for directions is limited, the residual encounters offer an opening to question Google's own incomplete and fragmentary representation.

Location awareness becomes the mechanism through which Google Maps diffuses its global mapping project, a benign consciousness set against the well-known exclusions in the act of mapping. While there are records of Google's corporate acquisitions process of buying and developing tools in the ongoing process of building a universal map, the nuts and bolts of the process are still mostly opaque. Google Maps provides little insight into how its algorithms work to search for directions and plan routes.[93] Google's safeguarding of its proprietary and commercial algorithms has been widely criticised for the ways by which the secrecy naturalises its information control and what it makes invisible in the process.[94] Google obscures its processes of mapping—like its algorithmic search logics—to seem naturalised and reified,[95] often erasing the human labour,[96] and the environmental footprint of Google's social and material mechanisms.[97] Instead, Google asserts its goals through carefully controlled online declarations, blackboxing how their systems work and obscuring lines of ownership through restructuring.[98] There is still much to glean from the little information that is shared and the discourses that help shape the so-called black box.[99] And there are many studies that follow this route of analysis through Google's limited public-facing records that investigate forms of "surveillance capitalism";[100] the perpetuation of oppressive and racist systems through algorithmic logics;[101] the reification of class divides perpetuated by Google's now-defunct Google Glass initiative;[102] and, Google's proprietary logic that will map public infrastructures like roads while also concealing their own data infrastructures like data centres. These all point to the ways that dissecting the actual code is not the only way to critique Google Maps—their discourses and their operations are also revealing.

Interrogating these imaginaries and the ontologies they reify is a practice reflected in Karin Fast and Pablo Abend's methodology of geomediatisation realism. For Fast and Abend, geomediatisation realism "entails the call to *de-center* the media by looking at the practices and operations surrounding geomedia rather than concentrating on the properties or functionalities of a set of discrete objects or technologies."[103] So while Google might not share the coding of their map, it reveals values encoded in its map and mapping software in other ways. The language and rhetoric Google and Google Maps employ are constituents of how the map presents and is present in the world. These discursive manoeuvres are built into and made visible through the Google Maps application interface, in the Google Maps "About" webpage, its official blog The Keyword, where product managers provide product updates, and in Google's product developers' presentations at their annual product conference, Google I/O. The rhetorical manoeuvres enmesh mundane forms of wayfinding in a technological imaginary of enhanced everyday life for an idealised "user" that helps position Google Maps as a public service and not a commercial platform, how it helps make space easy to understand and easy to navigate in its map flexibility, and how it produces reliable ordering of place. The maps promise to be an objective reference for public service, a neutral tool that makes space legible, a platform for a self-sufficient user who can explore anywhere, and an adaptable map, amenable to any technical glitches; but these are plaited with mapping practices that make the promises possible or at least believable.

## Location Awareness in the Age of Google Maps

*Location Awareness in the Age of Google Maps* proposes a critical language for discussing location awareness without lamenting a sense of direction surrendered to Google Maps. The chapters of this book run through four registers of Google Maps' location awareness directive: public good, self-sufficiency, legibility, and error. The first two chapters consider the *geographies* of a location awareness mediated by Google Maps, considering how publicness and individual claims to space are organised within Google Maps' visions of totality. The next two chapters take on the question of legibility and error to consider how Google's location awareness orients what a readable and precise representation of space looks and feels like, and the double bind of being on the map. Each chapter opens with a vignette from my experience of asking for directions. The retelling of these encounters anchors a starting point for the concept.

Chapter 2, "Geographies of Public Good," examines how Google Maps leverages itself as a public mapping service through its promise of a universal map that organises what counts as public while silmultaneously grafting onto public resources. The chapter begins by establishing how Google promises to map the world through its project Ground Truth. Ground Truth orchestrates multiple types of externally and internally collected mapping data from sources such as satellites, government survey maps, as well as Google's ambitious Street View Project. Street View is a central part of how Google builds out its map as well as how it performs and arranges publicness.[104] This chapter digs into one of Street View's public mapping projects facilitated through its partnership with Aclima, an air pollution sensor manufacturer based in San Francisco, California. In this project, Google Maps attaches pollution sensors to Google's Street View cars for a street-by-street picture of the city's air quality. This initiative, titled Project Air View, is used to "measure" and "analyse" as well as visualise these values on Google's Environmental Insights Explorer dashboard. Google plugs this data into Google Maps navigation function to suggest "cleaner" routes for travel. However, a closer visual analysis of Google's air quality dashboards for the cities of Oakland, CA and Houston, TX reveal that Google's data reinforce segregated ordering of place based on techniques of digital redlining. The comparison of maps demonstrates how Project Air View's maps reproduce racist measurements of "risk" which serve to further entrench discriminatory organisation of cities, effectively organising who is included in calls to publicness. Google's public calls for sustainability reflect what Berlant terms "individual acts of consumption and accumulation,"[105] and uphold what Max Liboiron argues are the discriminatory goals of pollution.[106]

From there the chapter considers Google's organisation of publicness at the site of the data centre. Here the chapter digs into Google practices of what Shannon Mattern terms a "grafting" onto other public resources.[107] The data centre manifests this tension in the imaginary of Google as a public resource—it organises which spaces are helped and harmed while it takes from the public and it calls for the public to invest in them. The limitations of Google Maps' publicness are animated through how the data centre exists on the maps and the ways ordinary people try to publicly talk back to Google Maps via the Google review system. In closing the chapter, I review how data centres *on* Google Maps become spaces to review Google *through* Google Maps—an ouroboros of public review. It looks at the ways that location awareness is both a tactic of sharing information and also a method of obscurement that is about hiding which areas count and which ones do not, according to the terms of the map.

Chapter 3, "Geographies of Self-Sufficiency," moves from public service to individual acumen mediated through Google Maps' location awareness. In this sense, location awareness becomes a means to organise space through individual claims to space. This chapter drills into exploring and experiencing the central promises of the map. Exploration and experience operate as technological prerogatives that absorb space in the name of control. This chapter considers how in organising space around the explorer prototype, Google Maps extends a colonial sanctioning of space. Here the chapter draws parallels with Jas Rault's analysis of *transparency* as communication technologies that leverage promises of accessing the truth in the name of entrenched colonial administration.[108] Rault's framework re-orients *exploration* as a tool of settler possession and entitlement. In this context, simultaneous to the promise of exploration is ongoing negotiations of space based on what Garnette Cadogan describes as personal and political "cognitive maps of safety and danger" organised around calculations and geographies of risk.[109] Google's coded depoliticisation of spatial mobility (it's for everyone! To go anywhere!) reifies mobility as a mode of power and exploration as the means to attain it. Building a whole mapping infra-structure around the positionality of unbridled access erases so many experiences and reinforces unjust enclosures of space.

Through an examination of how Google situates itself in a specific trajectory of mapping, to how it upholds a geography of entitlement to everywhere without considering the social and structural forces that make space differen-tially available and unavailable. Through affordances of mapping businesses and best experiences, the map orders space along a single axis of consumption. The chapter closes with an examination of interventions into Google's mapping. What these critical modes of access demonstrate are the hard limits to Google Maps' view of "the universal" and how geographically insufficient Google Maps is in the face of all the radical spatial practices beyond the map. It questions a location awareness premised on a self-sufficient prototype.

Chapter 4, "Orientations of Legibility," is an analysis of how Google Maps directs an imagination of space as a simple surface to be read and managed. The chapter traces some of the cracks in the Google Maps edifice of legibility by first considering and how mapping helps to stabilise and coalesce meaning. From there it moves to the project of legibility, which is a project of making space *appear* governable. The chapter considers the process of taming spaces in the context of other mapping precedents from the London Tube Map to Kevin Lynch's project of prototyping the imageable city.[110] Google Maps continues these processes through its project of location awareness that works to both template and contain space, subsuming the fantasies of what makes a city legible.

This chapter locates how the ease of navigation is wrapped up into these systems of abstraction in the name of legibility. It considers the tension of fixity and flux within Google's templating of legibility. Fixity draws on what Didem Özkul theorises as the imposition of the algorithmic fix,[111] while flux reflects what Nanna Verhoeff terms the "performative cartography" of navigating with screen-based interfaces that complicate the "visual regimes" of navigation.[112] The chapter considers this tension in light of a project based on asking directions and the types of spatial scores produced during these encounters. It then moves to Google's latest project "Immersive View" and its attempt to model ways of reading space whilst claiming new modes of legibility. In drawing attention to the punctures in Google legibility, the goal of this chapter is not to fill the holes but instead to stand in the fractured fantasy that space can be held.

Chapter 5, "Orientations of Error," examines Google Maps' conditions of accurate spatial representation based on what is present on and what is absent from the map. The chapter looks at erasure from the map. In 2008, Buffalo, New York residents noticed that Google Maps labelled their area Medical Park, referencing a series of development projects instead of their community.[113] It considers how Google manages presence and absence from the map and how these constitute mapping error and accuracy. Looking at Google Maps as a form of establishing which neighbourhoods count and which ones don't, this chapter considers how the slippery definitions of absence and presence, and reflects what Anna Lauren Hoffmann terms the "discursive violence"[114] of data inclusion tactics that are entangled in systems of digital coloniality and imperialist capitalism.[115]

The chapter closes with Google's formalised projects of inclusion, namely their project of Street View mapping the favelas of Brazil and their Plus Codes project, Google Maps' ongoing practice of spatial data collection that "gives addresses" to those who "don't have an address."[116] Drawing on the work of Toks Dele Oyedemi, the chapter closes with a reflecton on "digital inclusion" in the context of data colonialism.[117] I consider how Plus Codes centralises Google Maps as a core infrastructure of capital flows by giving people addresses that are only legible to the Google Maps platform. to help facilitate the expansion of the Google Maps project. Through an examination of Google's Plus Codes projects, this chapter considers the hidden costs of data inclusion and data legibility tactics entangled in systems of digital coloniality and imperialist capitalism.[118]

This book navigates the frictions of Google's mapping project delivered through promises of seamless location awareness. It considers the unequal distributions of mobility and fixity, public good and risky publics, and missed turns and missing places. Google Maps does more than just locate spatial information but *organises* location awareness as a standard rather than an ever-changing relation to space. As this book shows, the stakes of using Google Maps—and the stakes of Google Maps becoming such a dominant navigational tool—are not just about getting lost and being found. Rather, the stakes are about how Google Maps allocates value to space, making claims to it, in the name of constructing its universal map.

Google Maps' promises of publicness, legibility, self-sufficiency, and accuracy operate through stratification in the name of location awareness. Rather than a universal map, Google Maps builds what Leanne Betasamosake Simpson calls "incomplete worlds on incomplete knowledge."[119] The consequence, Simpson writes, is that "we risk relocating the very oppressions we are trying to liberate ourselves from."[120] It is not that a sense of direction is distorted via the lens of Google Maps, but that that Google Maps' location awareness is a distortion of publicness, access, legibility, and precision. But set against Google's information project of totalising location awareness are people refusing this universalising vision in profound and mundane ways, challenging Google Maps' renderings and positionings while also using Google according to one's own terms. This book sits in this tension in the name of expanding what location awareness in the age of Google Maps includes and what it takes for granted.

## Notes

1 Noone, "Locating Embodied Forms," 635–644; Noone, "Navigating the Threshold." From the art project *From Here To*. For more, see: https://www.thereroutingproject.org/from-here-to-rebecca-noone.
2 Along with Professor James Spilker, Hugo Fruehauf, and Richard Schwartz. Parkinson also won the Marconi Prize, another prestigious award for engineering, in 2016.
3 Myers, "Bradford Parkinson"; Carey, "Stanford Engineer Bradford Parkinson"; Olson,"The 'Father of GPS.'"
4 For a history of GPS technology, see Paul E. Ceruzzi's *GPS* and J. Lee's Global Positioning/GPS entry in *International Encyclopedia on Human Geography*. See also Goodchild and Janelle, *Spatially Integrated Social Science*. For a feminist analysis

of the Geographic Information System see Kwan, "Feminist Visualization: Re-envisioning GIS."

5  Whipple, "GPS Creator Bradford Parkinson." See also Parkinson's discussion of how he "really doesn't like having his location tracked" in Olson, "The Father of GPS."
6  Whipple, "GPS Creator Bradford Parkinson."
7  Whipple, "GPS Creator Bradford Parkinson."
8  Carey, "Stanford Engineer Bradford Parkinson."
9  Chow, "Age of the World Target"; Parks and Kaplan, *Age of Drone Warfare*; Virilio, *War and Cinema*.
10  Kaplan, "Precision Targets," 693–713.
11  McQuire, "One Map," 150–165.
12  Graham and Dittus, *Geographies of Digital Exclusion: Data and Inequity.*
13  McQuire, "One Map," 154.
14  Madrigal, "How Google Builds ITS Maps."
15  This is elaborated on in Chapter 2 with a discussion of Google Maps' Project Ground Truth. Also see McQuire, "One Map"; and Madrigal, "How Google Builds Its Maps."
16  Harris, *End of Absence.*
17  Petrow, "I Was a GPS Zombie."
18  Petrow, "I Was a GPS Zombie."
19  Romm, "Using Google Maps."
20  Milner, "Death by GPS?"
21  Plantin, "Google Maps as Cartographic Infrastructure," 489–506; Plantin et al., "Infrastructure Studies Meets Platform Studies," 293–310.
22  Farman, *Mobile Interface Theory*; Farman, "Map Interfaces."
23  Farman, "Map Interfaces," 88; drawing on Casey, *Getting Back into Place,* 36–37.
24  Lloyd, *Information Literacy Landscapes*; Lloyd, "An Emerging Picture," 570–583.
25  Lloyd, *Information Literacy Landscapes.*
26  Verhoeff, *Mobile Screens,* 134.
27  Galloway, *The Interface Effect.*
28  Massey, *For Space.*
29  Massey, *World City,* 22.
30  Massey, *Space, Place, and Gender,* 1994; Massey, *For Space.*
31  Tonkiss, *Space, the City, and Social Theory,* 23.
32  McKittrick, *Demonic Grounds,* 60.
33  McKittrick, *Demonic Grounds.*
34  For more on how maps produce knowledge, see: Harley, *New Nature of Maps*; Drucker, *Graphesis*; Eades, *Maps and Memes.*
35  McCullough, "Urbanism of Locative Media," 26–29.
36  Goggin and Hjorth, *The Question of Mobile Media,* 3–8.
37  Gordon and de Souza e Silva, "Net Locality." See also de Souza e Silva and Frith, "Locative Mobile Social Networks," 485–505.
38  Gordon and de Souza e Silva, "Net Locality," 13, my emphasis.
39  Gordon and de Souza e Silva, "Net Locality," 2, my emphasis.
40  Goggin and Hjorth, *The Question of Mobile Media,* 3–8.
41  de Souza e Silva and Frith, *Mobile Interfaces in Public Spaces,* 10.
42  Galloway and Ward, "Locative Media;" Tarkka, "Labours of Location: Acting in the Pervasive Media Space."
43  Frith and Özkul, "Mobile Media beyond Mobile Phones," 294; drawing from Özkul, "Location as a Sense of Place," 2015.
44  Halegoua, *The Digital City.*

45 Halegoua, *The Digital City*, 6.
46 Barns, *Platform Urbanism*.
47 Massey, *Space, Place, and Gender*.
48 Elwood, "Digital Geographies," 221.
49 Walcott, *On Property*; Fiske, "Surveilling the City."
50 Waidner, *Sterling Karat Gold*.
51 Hawkins, "Deep Dive into Google Maps."
52 As argued in McQuire, "One Map" and *Geomedia*; Rowan Wilken, "The Business of Maps," and Graham and Dittus, Geographies of Digital Exclusion.
53 McQuire, "One Map," 154.
54 Gannes, "Ten Years of Google Maps, From Slashdot to Ground Truth: Ten episodes from the dawning days of Google Maps."
55 Bret Taylor Quoted in Gannes, "Ten Years of Google Maps."
56 Gentzel, Wimmer, and Schlagowski, "Doing Google Maps," 151–152.
57 Muehlenhaus, *Web Cartography*.
58 Cowley, "Google Snaps Up Keyhole," *InfoWorld*.
59 Google, "Keyhole Markup Language."
60 Muehlenhaus, *Web Cartography*; Presner, Shepard, and Kawano, *Hypercities*.
61 Chowdhry, "History of Google Acquisitions"; Wilken, "The Business of Maps."
62 John Markoff, "That's the Weather."
63 Graham and Dittus, *Geographies of Digital Exclusion*.
64 Joël Kalmanowicz on Google for Developers, "Google Maps APIs."
65 Presner, Shepard, and Kawano, *Hypercities*.
66 Graham and Dittus, *Data and Inequity*; McQuire, "One Map."
67 Gordon and de Souza e Silva, *Net Locality*, 20.
68 Plantin, "Google Maps as Cartographic Infrastructure."
69 Loukissas, *All Data Are Local*.
70 See also Presner, Shepard, and Kawano, *Hypercities*; Shekhar and Vold, "Geographic Information Systems," 91–125; Bray, *You Are Here*.
71 Thiagarajan and Akasaka, "Building a Map for Everyone."
72 Google Maps, "My Maps."
73 Kitchin et al., "Conceptualising Mapping," 6.
74 Strauss, "The Imaginary," 330.
75 Suchman, "Anthropological Relocations," 3.
76 Bucher, "The Algorithmic Imaginary," 39–40.
77 Anderson, *Imagined Communities*.
78 Berlant, *Cruel Optimism*, 11.
79 Berlant, *Cruel Optimism*, 11.
80 Irani, *Chasing Innovation*.
81 Benjamin, *Race After Technology*.
82 Nakamura, "Feeling Good About Feeling Bad," 47–64.
83 Massey, *World City*, 88.
84 Ahmed, *Queer Phenomenology*.
85 Ahmed, *Queer Phenomenology*, 3.
86 Reid, "How 15 Years of mapping the world makes Search better."
87 Brouwn, *This Way Brouwn*.
88 As quoted in van der Meijden, *This Way Bruown*, 96.
89 Lehmann, "Stanley Brouwn," 60.
90 Debord, "Construction of Situations," 29–50; Debord, "One Step Back," 25–28.
91 Tonkiss, *Space, the City, and Social Theory*, 100.
92 Filliou, *Teaching and Learning*.
93 Vaidhyanathan, *The Googlization of Everything*; McQuire, *Geomedia*; Noble, *Algorithms of Oppression*.

94 Lucas D. Introna and Helen Nissenbaum. "Shaping the Web: Why the Politics of Search Engines Matters." : 169–185;
   Noble, *Algorithms of Oppression.*; Vaidhyanathan, *The Googlization of Everything.*
95 Noble, *Algorithms of Oppression.*
96 Gillespie, *Custodians of the Internet.*
97 Holt and Vonderau, "Where the Internet Lives," 71–93.
98 Zuboff, *The Age of Surveillance Capitalism.*
99 Bucher, "The Algorithmic Imaginary," 30–44.
100 Zuboff, *The Age of Surveillance Capitalism.*
101 For examples see Benjamin, *Race after Technology*; Gilliard, "Friction Free Racism"; Noble, *Algorithms of Oppression*; Safransky, "Geographies of Algorithmic Violence."
102 Noble and Roberts, "Through Google-Colored Glass(es)," 187–212.
103 Fast and Abend, "Introduction to Geomedia Histories," 2387.
104 McQuire, "One Map"; Plantin, "Google Maps as Cartographic Infrastructure."
105 Berlant, "The Face of America," 178.
106 Liboiron, *Pollution Is Colonialism.*
107 Mattern, *The City Is Not a Computer.*
108 Rault, "Window Walls and Other Tricks of Transparency."
109 Cadogan, "Black and blue."
110 Lynch, *Image of the City.*
111 Özkul, "Algorithmic Fix."
112 Verhoeff, *Mobile Screens.*
113 Dewey, "How Google's Bad Data Wiped a Neighborhood off the Map."
114 Hoffmann, "Terms of Inclusion."
115 Mervyn et al., "Digital Inclusion and Social Inclusion," 1086–1104; Milan and Treré, "Big Data," 319–335; Gangadharan, "Downside of Digital Inclusion," 597–615; Oyedemi, "Digital Coloniality," 329–343.
116 Google Maps, "Learn about Plus Codes."
117 Oyedemi, "Digital Coloniality and 'Next Billion Users.'"
118 Mervyn, Simon, and Allen, "Digital Inclusion and Social Inclusion" and Milan and Treré, "Big Data from the South(s)."
119 Simpson and Maynard, *Rehearsals for Living*, 292.
120 Simpson and Maynard, *Rehearsals for Living*, 292–293.

# 2 Geographies of Public Good

**Organising | Grafting**

In my circuit through central London, I found myself standing near the north end of Millennium Bridge looking for directions to the British Museum. The person I approached thought about it for a moment, then offered: *I think it's by King's Cross station.* With this announcement, they drew a straight line across the piece of paper I offered and suggested I take the London Underground to get there since we were close to St. Paul's station. First, they told me to head westbound on the Central Line from St. Paul's. Then, they paused and told me to go eastbound. Then wait, no—westbound. After volleying back and forth, they pulled from their bag an Official London Underground Map—a paper map—verifying that I was indeed to head west on the Central Line. I was about to set off, when they decided to triple-check which direction I should go, this time using Google Maps. *Just to make sure*, they said.

Opening Google Maps, my director saw that they were sending me off in the wrong direction (a point likely already flagged by some of you). Firstly, King's Cross is not on the Underground's Central Line. And, as some might have also noted, King's Cross is not the closest stop to the British Museum (but also not the worst option). Seeing this, the person read aloud the new set of instructions from their phone. They told me that I should head westbound on the Central Line and get off at Tottenham Court Road station, and not King's Cross, after all. I was to walk from Tottenham Court Road station, which is, indeed, on the Central Line. After translating this new set of directions from their phone, they proclaimed: thank goodness for Google![1]

"Thank goodness for Google" was a passing comment; but it nevertheless resonated with the similar hyperbolic and jokey expressions animating my other experiences of asking for directions. Utterances of (something to the effect of) "thank goodness for Google" punctuated many encounters—often offered in solidarity with the circumstances of being lost, paired with an appreciation for the convenience of Google's omniscient location awareness. While calling on Google Maps for directions does not mean a relinquishing of wayfinding to Google Maps nor a complete dependency at the expense of a "sense of direction," the plasticity of "thank goodness for Google" struck a chord.

DOI: 10.4324/9781003251569-2

While a playful affectation, Google Maps as a public service is an imaginary that Google actively promotes. Google's oft-repeated founding mission is to organise the world's information and make it "universally accessible and useful;" a promise it bestows upon its products like Google Maps.[2] As Mark Graham and Martin Dittus establish at the start of *Geographies of Digital Exclusion,* Google Maps is among one of the largest "publicly available" platforms of aggregated geographic knowledge, scaling the vastness of the whole earth image to the immediacy of the everyday.[3] Foundational to Google Maps' claims to public service is its promise of building and sharing a universal map: the one map that can be freely used by everyone, everywhere. It is "a bird's-eye view of Earth, from the highest mountains to the lowest valleys and everywhere in between,"[4] paired with its promise to "bring helpful local information to your fingertips."[5]

While Google often obscures the specific details of how it adds and processes information,[6] projects like Google Street View enact a type of public performance of mapping. As Scott McQuire has argued,[7] Street View is central to Google's appeal to publicness as well as its paradoxical pursuit of being, in McQuire's terms, "one map to rule them all."[8] The car driving around the world is Google's visible performance of documenting everyday pathways for its Street View map—a means to be seen *seeing* in its pursuit of mapping everywhere. With its on-the-ground precision of vision, Street View is there to supplement other parts of Google's mapping processes such as its satellite imagery. The scope of an entire planet coded from above and from below in turn predicates an objectivity and a totality of representation made possible through Google's technical and logistical prowess to traverse these scales. At the same time, Google's performance of mapping everywhere enables it to both define what counts as *everywhere* in its project of mapping the globe, and to access and lay claim to that *everywhere.*

Through its ostensible hold on completeness, Google Maps leverages itself as an invaluable resource for the global public. As Google expands mapping to include the mapping of air quality (to help people find "more sustainable routes" or "helping cities plan for the future"), Google Maps entrenches itself as a flexible tool for "helping out" amidst global catastrophes. Google's contortions of publicness reflect what Lauren Berlant describes as the "production of mass political experience,"[9] played out through "scenes of private acts."[10] Rather than providing meaningful insights for public use, Google Maps' classification of spatial information in fact works to *organise* publicness: it does not fundamentally mitigate risk but, instead, spatially distributes risk and safety across populations, often along existing power formations.

### Organising Space: Ground Truth and Street View and Performances of Objectivity

Google's founding mission (to *organise the world's information and make it universally accessible and useful*) is a familiar claim of mapping projects. At the same time, connecting the organisation of the world's spatial information to the everyday applications (its access and use) is no small claim. Indeed, the scale of Google's mapping project is reflected in its name Project Ground Truth. Project Ground Truth brings together "in-house" and "external" spatial data from satellites, GPS data, historical maps, government and state sources, and, significantly, the information processed through their ambitious Street View project.[11] Google's Ground Truth stitches together of this patchwork of spatial information to produce a totalising image of the world augmented by layers of data.

The evocative name Ground Truth predates Google's mapping project. One etymology of the term pins it to Henry Ellison's poem "The Siberian Exile's Tale" (c. 1833),[12] where Ellison conjures the language of *ground truth* to signify a divine knowing—at once fundamental and otherworldly. Google Maps adopted the term "Ground Truth" from the U.S. Department of Defence's internationally used Geographic Information Systems (GIS) and the military language of remote sensing. Here, ground truth describes a relationship between territory and data that produces omniscience.[13] The top-down perspectives of GIS and remote sensing enable a simultaneous multi-directional *seeing through space*, and the precision of identifying a fixed location, to arrive at a ground truth.[14] Important to underline here is how this totality is in the service of militaristic tactics and imperial dominance with the logic that to know all is to own and control.[15] As Lisa Parks and Caren Kaplan have argued, the applications of ground truth as part of the deadly U.S. military invasions of Iraq in the 1990s and again in the 2000s, betray the imperialist underpinnings of ground truth ontologies.[16] Moreover, Kaplan contends that ground truth's military context of aerial assaults and drone warfare, produces "target subjects."[17] In this sense, GIS and its promise of ground truth are, in Haraway's terms, both "the eye that fucks the world" and a "God Trick" wed to fantasies of military dominance based on the singular and totalising view from above.[18]

In feminist geography, Susan Roberts and Richard Schein, and Liz Bondi and Mona Domosh frame GIS as a masculinist technology premised on the power of the gaze propelled by a desire of looking.[19] Mei-Po Kwan and others argue that feminist epistemologies of positionality are incompatible with GIS positivism and its singular gaze.[20] In pushing against the belief in ground truth's precision as the proximate of truth, feminist epistemologies contend that space is relational and subjective, and as such spatial meanings cannot be measured.[21] And while many authors have contributed to dispelling the fantasies of objectivity sutured into total world visioning at the hands of militarised and state-run mapping systems like GIS,[22] Google's view from above is nonetheless key to Google's purchase on publicness. This scaling of a global network of information to the individual is part of how Ground Truth gets taken up as a public service. Google Maps makes ground truth portable in a way that enables one to carry this *truth* around in one's pocket for the sake of ordinary pleasures, curiosities, and getting things done, translating the menacing precision of warfare ground truth to the logistics of everyday life.[23]

Google's project Ground Truth promises a total and objective vision of the world—a positivist assumption made in the name of military dominance, claims to space, and the preservation of power. John Pickles argues that this is the "depicting of the world from above not only suggests the sophisticated space-age equipment used in, say, a global positioning system; it also claims the entire world as a potential data source."[24] This is the continuum of ground truth along which Google Maps tracks. As Jason Farman has noted, maps like Google Maps and Google Earth often evade such scrutinising of their power and their omissions because their use of multiple informaton sources, such as satellites and aerial photography, seemingly captures everything.[25] Moreover, the "total" view is made mundane when popularly applied as an everyday wayfinding tool.

Google Maps promises a system of mapping that is not only total, but, *on the ground*—a system that is continuous, that captures detail and the big picture, and that keeps pace with change. To do this, Brian McClendon, former head of Keyhole and VP at Google Maps and the person credited with project Ground Truth, explains that there needs to be not just an assemblage of licensed third-party data from government maps, satellite images, and regional surveys, but also in-house data collection and processing. In response, Google developed the Street View project, a central part of Google's Ground Truth.

### Imagining a Public Vision

A central part of Google's Ground Truth project is Google's Street View project. Street View allows one to see images of a place from the level of the street. Street View's data collection process hinges on a single car (now multiple cars in distinct parts of the world) driving one road at a time while taking a non-stop series of photos, in 360-degree horizontal and 290-degree vertical orientations. These images are then stitched together to produce a continuous sequence of spatial imaging, or what Scott McQuire terms a "montage." McQuire writes, "The fact that each panorama displayed in the Street View application is composed from a variety of separate 'shots' effectively embeds spatial multiplicity and temporal duration into what is presented as a continuous picture. This embedding means that 'continuity' in Street View translates to a "seamless montage of the world from the vantage point of the road."[26] Street View offers a seamless pictorial view of the world from the perspective of an intrepid automobile taking on the world's roadways.

Street View is Google's public performance of data collection. The Street View car is the visible and material presence of Google at work. In its design, Street View appears like a fun, low-tech science project—a car, overlaid in graphics, with a strange spherical camera perched on its roof—something that may have emerged from the same now-mythical garage where Larry Page and Sergey Brin created Google's Search Engine. The Street View project is the fantasy of one small car tracking the world—a streamlining of the globalised collection to a singular intrepid vehicle surmounting the logistical complexities of driving every road. The project manifests in the ultimate Americana fantasy of claiming space through the open road and the moral project of being a good and informed driver[27]—an imagination of total freedom, autonomy, and mobility translated through the presumed objectivity of the photographic image. The sheer scale of recording every road around the globe is alluring and ambitious, and at the same time delimits the type of spatial information that can be recorded to that which can be seen from the street (now expanded to include the passages, pathways, and interiors tracked via the on-foot Street View Trekker).[28]

In 2022, Google celebrated the 15-year anniversary of Street View. Over those 15 years, Google claims that its Street View car(s) "circled the planet 400 times,"[29] or approximately 26 times per year, or over twice per month for 15 years. By taking this endless roadtrip, Google performs the inconvenient task of documenting every street so we can drive to work with ease. To mark the anniversary, Google and Wired magazine produced a video titled "All the Ways Google Gets Street View Images," narrated by Ethan Russell, a Senior Developer at Google Maps. In this video, Russell recounts the history of Google Maps' technological development, outlining the early promises of digital mapping and the tools Google developed to advance these promises. In the video, Russell lauds the Street View's vision of taking photos of everything which Google "stitches" together into a "seamless experience that can let people explore the world." Street View is no longer just for cars but now mounts its cameras on bicycles, snowmobiles, backpacks, trolleys, and even sheep in the Faroe Islands[30] through what it calls the Street View Trekker. These new actors don cameras as backpacks to document terrain otherwise inaccessible by car like the favelas of Brazil[31] or the rockfaces of El Capitan in Yosemite.[32] More than just a means to access images of space beyond the street system, the Street View Trekker, the wanderer accompanied by a robotic friend in the backpack—a 360-degree camera perched out of the backpack and peering over one's head is more intimate than the Street View car and leaves open the potential to go beyond the exteriority of streets, and map interiors (Figure 2.1).

According to Russell, the expansion of the Street View project opens up new spaces to Google's guided tour, stating, "We'd like to take you inside transit stations, or maybe even inside a museum, or out into the wilderness, to the top of Machu Picchu, and the top of Mont Blanc, down into the Amazon rainforests, and underwater by the Great Barrier Reef, or up in the International Space Station." Russell's pitch leverages an imagination of an impossible feat scaled to the immediacy of everyday convenience: Street View's frequency of collection, organisation of data, and the ever-expanding expectation of what is mappable translates to Google's goal to "take you" places through their technological expertise. Google Maps describes Street View as a "virtual representation of our surroundings" that "enables people

*Figure 2.1* Illustration of Google Maps' Street View car and Street View Trekker. Illustration by Colin Medley.

everywhere to virtually explore the world." While Street View presents as a pleasurable resource for armchair explorers (that virtual experience of the world!), these images are also the resources by which Google facilitates its claim to being the public's go-to, comprehensive, objective map via Project Ground Truth.

In 2012, Alexis Madrigal at *The Atlantic* published the article "How Google Builds Its Maps—and What It Means for the Future of Everything."[33] In the article, Madrigal speaks with Brian McClendon, Google Maps VP, and Manik Gupta, senior product manager at Google Maps about the work and protocols embedded in Google's claims to "objectivity" on which its status as the public's go-to map rests. Significantly, this article is one of Google's earliest public acknowledgements of its Project Ground Truth. Madrigal goes so far as to introduce Ground Truth as "the secretive program to build the world's best accurate maps."[34]

The following year, at the Google Maps I/O 2013, Michael Weiss-Malik, Engineer Director and Andrew Lookingbill, then-Software Engineer presented "Project Ground Truth: Accurate Maps Via Algorithms and Elbow Grease."[35] Then in 2014, Greg Miller at Wired Magazine published his article, "The Huge, Unseen Operation Behind the Accuracy of Google Maps."[36] The articles and the I/O presentation all claim to provide sneak peeks into the processes of collecting information. In lifting the curtain, these accounts reveal the centrality Street View plays in the entire Google Maps venture and the hidden orchestrations of computer vision algorithms and "meticulous manual labour" that process these images. This mix of software and human labour extract details like traffic signs, street names, lane widths, building numbers, and signage and input them on the map.

In his article, Madrigal interviews Nick Volmar, one of the so-called manual data inputters at Google Maps. As a *Ground Truth Operator*, Volmar is tasked with zooming into the satellite and Street View images to assess the layout of space. Operators like Volmar "extract" and add place-based information to the map, "coding every bit of logic of the road onto a representation of the world."[37] They also verify location information such as which direction one can turn at an intersection based on if there is a sign that marks no left turns or a one-way street. Operators also check for business addresses, speed limits, and respond to complaints filed by Google Maps users. Weiss-Malik and Lookingbill refer to the operators as the "elbow grease" of Google Maps, producing objectivity from the images outputs of the Street Views cars.

Street View's public presentation of collecting totality, reflects mechanisms of standardisation that Dylan Mulvin terms the *theatre of objectivity*: "the rituals that go into making systems appear objective."[38] The sight of the Street View Car driving past us on the street, collecting data about our surround, activates an imporant part of the performance. Another important part of this theatre is the deus ex machina (or elbow grease) of data processing and data management that translates Street View images to data points. Google does not share how many of these operators they employ or where they are based; but both articles speculate that this work spans the globe, reflective of the invisible and dispersed labour behind other platforms.[39] The scale of the workforce is implied through Google's claims to 99% coverage of the world (reliable, comprehensive data for over 200 countries and territories) with 25 million updates daily. Google's Street View car and its tireless deed of driving everywhere (the solo performer of sophisticated data orchestrations) ultimately works to naturalise its information control.[40]

Another familiar public protocol of Ground Truth that we participate in is CAPTCHA. In 2009, Google purchased CAPTCHA (which stands for Completely Automated Public Turing test to tell Computers and Humans Apart) and has integrated it into its information verification infrastructure across its many projects and platforms.[41] CAPTCHA is the process by which users attempt to prove to the computer that they are indeed human and do in fact want to make that credit card purchase. It calls on people to verify Street View classification, such as selecting images from a set that contains "traffic lights" or "motorcycles." CAPTCHA is part of a move to crowdsource the annotation of images from Google's various scanning projects, including Google Street View. We might, for example, be called to transcribe a house number that will eventually be used to update Google Maps. CAPTCHA is the blend of visible and invisible protocols of street view in which we are visibly called to participate and yet Google obfuscates how they use that data. As such, the public is brought into the piecemeal image processing of Ground Truth as well as the conceptual imagining of its totality.

The way Google collects, distributes, and circulates its information means that the public is caught up in Google's systems regardless of whether one uses the platform. Moreover, Google Maps profits from the public through CAPTCHA. Google deploys CAPTCHA to collect spatial information (from identifying street names to recognising pedestrian crossings to recognising bicycles), process public input, and licence that data to sell or use it for their own training of self-driving cars.[42] The public *annotates* Google's Street View imagery and verifies map information.[43] More than just a go-to-map, Google Maps is now what Jean-Christophe Plantin terms "cartographic infrastructure."[44] As Plantin argues, Google Maps' forms of data input thus produce an enclosure of online information access where one is brought into participating in its logistical imaginary of the total map even if one is not looking for directions.[45] Google maintains its authoritative position as being for the public with little attention to how the public works to maintain that mapping authority. Plantin shows how public participation is key to Google's cartographic infrastructure of Ground Truth "from participatory mapmaking to database maintenance," ultimately centralises participatory information practices.

Google's performance of objectivity through visible processes of the Street View car and CAPTCHA are staged within what Deborah Cowen terms "logistics space."[46] Logistics space is a way of conceptualising space not simply as territory but as the territory on which logistics occurs. Google Maps' process of collecting and maintaining spatial data for its totalising database—cars driving every road, satellites circulating the globe, and inputters adding details around the clock—demonstrates how Google Maps is produced through a logistical imaginary of knowing all space. The logistics are these impossible feats of driving, flying, around-the-clock adding that in turn produce what Hockenberry, Starosielski, and Zieger call a "global operating system" which they characterise as a "set of conditions for the circulation of information and culture."[47] The processes beyond this performance of out-of-view, shared only in press briefings. Instead, they present simply as the image we see in street view, a public resource for us to see the representation of space that we can use to help us imagine a space. The execution of these logistics is the performance of objectivity that supports this idea of a single, universal, total map as a public service from which to extrapolate other spatial truths.

A central part of Google Maps' promise of organising the world's information is its convenience of having all that information available anywhere. Through projects like Street View and the umbrella project Ground Truth, we see how Google Maps pairs their herculean feat of documenting the entire world, with the everyday convenience of checking the map for simple directions. But it is also important to remember that Google renders invisible the processes of continuous human labour (adding, checking, verifying mapping data) that make Ground Truth possible. While Google Maps appears an immaterial feat of data processing, rendered through extraplanetary satellites and an intrepid automobile, a closer look at Ground Truth uncovers a logistical network of people doing the work of walking, driving, and manually inputting who are part of this performance of data collection. These mundane practices of getting from A to B are ensnared in a logistical imagination that sells a fantasy that all life can be improved if we just add and share more data.

**Organising Safety and Risk**

Street View helps Google Maps mobilise its status as the comprehensive, objective, go-to digital map to position itself as a public service—one that can provide insights that will benefit the public beyond everyday navigation. For example, in Finland, Google Maps promises to "revolutionise" and "stream-line" public works projects like road maintenance. Here the Street View car drives through the northern roads of Finland, outside the main city centres, in search of potholes or cracks that need filling. The car is used to identify where maintenance is needed but does not perform the task itself. And while Google's advertised example of public service does not detail the processes of what happens next once a pothole is identified (is there maintenance beyond the monitoring?), what we see is how Google Maps' ubiquitous data collection practice also results in an auxiliary collection of maintenance data with its camera. It marks the risky potholes for governments to then proceed with road maintenance, in the name of making the roads safe. This monitoring of public space becomes a means of organising space according to terms of risk and safety. These terms of risk and safety are made to seem in support of public infrastructure.

One of Google Maps' promises is providing "environmental insights" through its maps tool[48] in the form of "clean air routing" and what it calls "eco-friendly routing."[49] These new features are part of Google Maps' Earth Outreach Program and the company's push toward the banner of "Sustainability" where it claims to be "accelerating climate action" as "part of public action of saving the planet."[50] Part of this is what Google calls its "vision for the future of Google Maps—an immersive, intuitive map that reimagines how you explore and navigate, while helping you make more sustainable choices."[51] These are sustainable decisions that according to Chris Phillip, VP and General Manager of Google's GEO project, help Maps feel more like the real world, with options like a "vibe check" and "stunning multi-dimensional views."[52] The result produces a mapping of "clean air" that ultimately classifies and indexes space according to terms of safety and risk.

One notable project in Google's sustainability initiatives is its AirView Project. Since 2015, Google's Street View program has partnered with Aclima, an air pollution sensor manufacturer based in San Francisco, USA. Aclima's sensors come with the promise of measuring air quality and what they call "hyperlocal pollution mapping"[53] made possible through the Street View's street-level spatial data collection processes. Together, Google and Aclima created a "specialised mobile air sensing platform" to develop

Environmental Insights Explorer (EIE) for Google's AirView Project. EIE air sensors claim to measure and analyse levels of pollutants such as nitrogen dioxide ($NO_2$), nitrous oxide (NO), carbon dioxide ($CO_2$), carbon monoxide (CO), fine particulate matter (PM2.5) and ozone (O3) in the air. Google Maps attaches the EIE sensors to Street View cars that drive around cities and collect pollution data, which it then translates to air quality maps. Yielding street-by-street picture of the city's "air quality," Google translates its insights to the EIE Insights Workspace dashboard. Here, over 3,000 cities can review and evaluate the emissions data Google collects and maps.[54]

Initially tested in San Francisco and Los Angeles, Google announced AirView's international expansion at the 2018 Copenhagen Climate Summit, promising partnerships with municipal governments around the world, making it "freely available" once one registers with the EIE system. Through data collection and processing, Google publicly shares this information to help people (who they categorise as "citizens, scientists, authorities and organisations") to "make more informed decisions and accelerate efforts in their transition to a healthier, more sustainable city."[55] AirView promises to help cities meet target transit emissions and understand which neighbourhoods are the most impacted by pollutants and to "help cities and governments translate these targets into concrete action" based on these street-level measurements. Except this data is activated via personal decision-making: the means to *choose* where to go based on air quality. It recommends, for example, that people "adjust their bike route or choose another time to exercise" based on the differences in air quality between certain areas or times.[56]

As part of AirView, Google claims it has taken 500 million air measurements to produce its dashboard maps, which create visual representations of cities according to pollution levels.[57] Neighbourhoods are marked by a gradient colour scheme running from dark red to light green to show areas with the highest level of pollution concentration. This dashboarding of the air quality sensory data is leveraged as a way to quickly know "where" the pollution is, and based on the logic of the promises above, *avoid* these areas marked as "unclean." In applying its mapping tools in this way, Google Maps effectively organises publicness through the logistics of individual choice. As I argue below, this makes air pollution a technical problem made solvable through individual action of personal investment, rather than as part of the structural discrimination undergirding an organisation of space via a logic of investment.[58]

The Google Earth Outreach website features two notable "case studies" in which Google partnered with the U.S.-based environmental organisation, the Environmental Defence Fund (EDF): Oakland, California, and Houston, Texas. According to the case study blurbs, Google selected Oakland and Houston because these cities contain some of the most polluted areas in the United States. For example, in the map of Oakland, South Prescott and Downtown Oakland are saturated with deep reds that signify high levels of toxicity. These are contrasted with yellow lines where there is less pollution. By selecting these cities, Google's Project AirView asserts itself as a type of public intervention—an infrastructure for providing the necessary data to prove these areas are polluted and to measure the concentration of toxicity.

A closer look at Google's air quality mapping of Oakland and Houston ruptures the fantasy that Google's collecting of air quality data is a public-facing project that will improve citizens' lives. More specifically, AirView's maps very clearly align with other segregationist mapping, namely the racist land valuation project of redlining. Redlining is a racist housing policy in America that systematically denied mortgages and home loans to Black and racialised people.[59] Federal Housing Association of America (FHAA) made redlining policy in 1942, formally restricting Black access to housing by way of refusing to insure mortgages. As Chris Gilliard and Hugh Culik explain, redlining was an investment practice based on "anti-Black racist prerogatives to protect white property."[60] The Federal Housing Association of America restricting access to housing by refusing to insure mortgage loans to any but white Americans. The FHAA established, formalised, and implemented a system of insurance based on racist measurements of "risky" and "safe" investments.[61] The Home Owners Loan Corporation created maps for American cities according to these metrics of risk, which as Gilliard argues, effectively "color-coded the areas where loans would be differentially available,"[62] where the risky areas were red or yellow (hence the name redlining) and the less risky areas blue and green. These redline maps as such were technologies generative of a racist geography that systematically entrenched these values of spatial inequality and legitimised practices of marking risk and security based on racist logics of capitalism and investment.

Overlaying Google's mapping of air quality onto the redlining maps reveals a similar geographic organisation of risk. In so doing, Google's air quality maps replicate the redlining maps' practice of formalising racist segregation. For example, suburbs like Clear Lake or Friendswood are light green on both Google's dashboarding of pollution and redlining maps, while the Fifth Ward is red on both redlining maps and Google's dashboards (see Figure 2.2). Google's mapping of air quality consolidates the risks of pollution with the racist evaluation of risks on housing investment, not only replicating the same racist logic but encoding this logic according to additional metrics of "risk." The AirView maps reify the legacies of racist housing policy that devalues these areas and leave them vulnerable to racist zoning and predatory industrial development (more on this in Chapter 5). Through these maps we see what the racist term "public service" is actually based on, and how, according to John Fiske, when "public" is instrumenta-lised, it becomes "whitened."[63]

Who lives in these neighbourhoods is important. A recent report by the Othering and Belonging Institute at UC Berkley shows that Oakland is among the most racially segregated cities in America (in the 92nd percentile).[64] The restrictive land use policies, racist housing policies, and histories of redlining are among the reasons for that.[65] For example, in Houston, West University Place, an affluent neighbourhood in the Houston area with a median income of

*Figure 2.2* Composite of AirView Dashboard of Houston with the Redlining Map of Houston Texas.

$190,000, is light green, the Sunnyside neighbourhood with a median income of $27,500 is dark red. According to Rice University's Kinder Institute for Urban Research, the Sunnyside neighbourhood is 79.4% Black and, in contrast, West University Place is 71.2% white.[66] This pattern is repeated across both cities. The Oakland and Houston "case studies" thus reveal how Google's project not only produces neoliberal assumptions about well-being in cities but reproduces uneven relations to space and reinforces the racist structures by which cities are organised (Figure 2.2).

Moreover, it is no accident that these neighbourhoods read as highly polluted on the AirView Dashboard. In Houston the dark red, "polluted" neighbourhoods are bookended or even surrounded by industrial plants emitting toxins into the atmosphere. This translates to higher cancer rates, unbreathable air, risks of asthma, heart disease, and premature death. West Oakland is wedged between three major highways. The neighbourhood is home to two major industrial polluters, namely the Union Pacific railyard and the Port of Oakland; it also hosts a waste transfer facility, a concrete plant, and a wastewater treatment plant.[67] The Port alone fills the neighbourhood with diesel emissions thanks to the ships but also the 9,000 trucks that drive to the port each year. A 2003 report conducted by the West Oakland Environmental Indicators Project (WOEIP), a partnership between Oakland-based non-profit Pacific Institute and the West Oakland community found that West Oakland has 90 times more diesel pollution per square mile than the rest of California.[68] Further, there is seven times more diesel exhaust per person in West Oakland than in the rest of its county.[69]

Google's project of abstracting pollution into dashboards—a mapping that ultimately classifies areas and neighbourhoods as risky or safe—only magnifies a city's prevailing inequitable arrangements. Google promises the precision to track down the pollution and identify where it is in the name of safety, but it treats these polluted areas as somehow naturally or unavoidably more polluted than other areas. According to AirView, these areas are simply polluted—a marking of space that does not intervene in what makes these areas polluted in the first place, from the deregulation of industry to the racist forms of city planning that create unjust proximities between residential and industrial areas.[70]

Google's mapping of air quality data reinforces and reproduces a segregated ordering of place based on techniques of what both Sara Safransky and Chris Gilliard "digital redlining."[71] Digital redlining is a process of engaging in racist classification schemes through opaque datafication systems.[72] This correlation to redlining maps evidence how Air Quality mapping reproduces the same racist logic of risk, amplifying the racialised valuing of space as safe and unsafe, or desirable and undesirable.[73] The performance of visualising toxic particles reinforces the redlining logic of risk, further racialising pollution. Therefore, exposure to pollution is more than simply a choice based on following a route on a map but something that is socially determined.

Strategies of valuing space as investable carry this racist baggage of the redlining prototype. Safransky's research on Smart City developments in Detroit.[74] Here real estate metrics are now based on something called Market Value Analysis (MVA) which has been used as a guide for the City of Detroit for the distribution of resources and development funds. Safransky shows how MVA is used as a dianostic algorithmic system premised on the reinterpretation of old "property regimes" that continue to overvalue white segregated neighbourhoods. As Sarah Elwood argues, "These digital media-tions re-inscribe racial logics that root personhood itself in ownership, autonomy, and individualism, in ways that rest upon the production and exclusion of a racialised other—and the myriad techniques of control/removal that have stabilised these relations over centuries."[75]

Google's data-driven systems of pollution management treats pollution as an evenly distributed threat—it is a stable substance that can be avoided. However, its mapping of pollution is a visual rendering of pollution's unevenness, reflecting the geographies of racism, colonialism, and classism. As Max Liboiron argues "unevenness is not just an accomplishment of pollution. It is its goal."[76] The colonial assumption that air can be measured is not to solve the inequity of pollution but to organise the public realm accordingly.[77] We see how *the measurable* is braided with the pursuit of sustainability,[78] and sold as what Susan Oman identifies as well-being data used to make "pivotal decisions" in the name of increasing life chances.[79] Here, where one lives is leveraged as an individual choice where one either chooses to live in a safe area or a risky area. This so-called choice is mediated via one's engagement with Google Maps. In the context of Information Literacy, Alison Hicks argues, "The assumption that an 'at-risk' status is both caused by and resolved through individual control also frames information literacy as a means of self-governance wherein people are expected to minimise risk through a continual pursuing of self-improvement and betterment."[80] So, while collection and sharing of air quality data are operationalised to confirm a cohesiveness to publicness or to the social, the dashboarding to make "sustainable choices" not only amplifies but also legitimises these orderings of space. Indeed, Google's publicness, or its presumed universalism, in many ways underlines the real and discursive violence that was there all along.[81]

Google's claims of "data-driven climate action" are undermined by the fact that this data already exists with community partnerships like that between the Pacific Institute, the Oakland Climate Action Coalition (OCAC), the OCAC's Resilience and Adaptation Subcommittee, Communities for a Better Environment, and the California Food and Justice Coalition, among others.[82] These groups have both authored reports on the pollution and emissions experienced by people in the community, and worked to support community members. For example, the OCAC is "a cross-sector coalition dedicated to racial and economic justice that engages Oakland residents in creating and implementing climate solutions that strengthen the environment, economic, and social resilience of frontline communities."[83] These efforts are directly tied to issues of food justice and land access, transportation and land use, actions based around community care and advocacy that underline the limited scope of Google's re-routing plans.

Google's AirView mapping presumes that pollution itself is a static substance that can be mapped. Beyond the science of this, Google Project AirView is based on the assumption that individuals can avoid pollutants by moving around them, choosing to live somewhere else. According to Google's directives of how to use its information, for example, West Oakland is an area to be avoided. But that advice leaves out the reality that many people live there, their families live there, they have histories there; and, therefore, people cannot simply "avoid" this area in order to comply with a map. To present space as open to individual choice—the option to go somewhere else, choose to live somewhere else, buy property somewhere else, invest in somewhere else—conditions relations to space according to imaginaires of untethered mobility. Shared as simple measurements, Google evades the social dimensions of pollution and instead offers repackaged and rebranded air quality measurements to replace data that these communities already have. Yet Google's insistence on measuring pollution in this area in the name of choosing safe places to live only enforces a belief that pollution in these areas is inevitable. The result organises location according to terms of desirability, further abstracting the root cause of the problem, and making pollution itself seem so inevitable as to foreclose the possibility of change or resistence.

Moreover, these measurements of risk are not static representations of place: they travel and accumulate new ways of valuing space. This circulation of spatial data is furhter evidence of what Plantin terms the "cartographic infrastructure" of Google Maps. As the work of Yanni Alexander Loukissas has shown, "local" map data gathered from one source recirculates through other popular location-based platforms like Zillow and Airbnb. Specifically thinking of Zillow, Loukissas tracks platforms that allocate value to location, measured through home pricing and rental costs in the case of Zillow, a real-estate platform, or tourism, pricing, and investment, in the case of Airbnb, these other sources of data get woven in and part of the algorithmic procedures of assessing housing value.[84] Moreover, they can become baked into government data, as recounted in Safransky's research on smart city mapping in Detroit.[85] Air Quality data becomes part of making decisions about where to live and where to go beyond Google Maps, naturalising these classifications of space and reinforcing racist valuing practices. And in making spaces not investible, or places to avoid, they become deficient of other social infrastructure to support the space.[86] Instead it's the tools that evaluate risk that become investible.

Investment in these geographies of air quality risk and safety operates on the large scale of global trade and industrial processing, but it also works at the level of commercial retail. For example, since 2013, Google has partnered with Breezometer, a company that visualises and processes location-based air pollution "at the neighbourhood level" through Google's Cloud Platform.[87] Breezometer recently announced a partnership with L'Oréal, the makeup and skincare company, to "uncover new insights around how the environment affects skin aging, and ultimately provide new services to consumers that can accompany their skin needs all over the world with personalised routines and lifestyle advice."[88] The air quality data is used to sell skin creams to off-set the "damage" of air pollution. Through Breezometer we get a sense of how this type of air quality information is further commoditised. Here the marketing of skin care products is put in stark contrast with the effects of pollution in places like West Oakland where low-income people are facing the challenge of breathing, a very real threat that does not have a simple product to simply buy and apply. Rather than connect and work with local activism already taking place to address not just pollution but racialised exposure to pollution, AirView propels private, individual responses to public problems. We see how Google Maps is organising rather than measuring pollution or contributing to the city's capacity to deal with it as this data becomes a means for marketing.

Google's project Ground Truth takes a partial view of the world and extrapolates totality. And while Google makes visible its claims to innovations it hides the human dimension of its data processing and the global economic conditions that make the scale of this data processing possible. Google's position as a public service through its classification of safety and risk only serves to reinforce and naturalise the inequitable spatial relations that already constitute city life. These are the geographies of publicness Google Maps enforces through its logistics of mapping. Google Maps does not actually contribute to public action but organises the city based on pollution levels to measure what is a "good" neighbourhood to live in and what is a "bad" neighbourhood to live in, measuring where has value and where does not. Google Maps is *organising* publics rather than measuring pollution or contributing to the city's capacity to deal with it. Instead, it frames pollution as another metric for computational tools to measure, in the name of advancing the total vision of space, and in doing so abstracts the root causes of what makes space unevenly risky in the first place.

### Grafting Publics

AirView exemplifies Google Maps' position as a public resource of objective data, leveraging an open-ended promise to provide insight into "collective" challenges like environmental pollution. From projects like Street View and AirView, we see how "the street" and "street-level" data play a central role in Google's claims to publicness, from routing to mapping pollution measurements. Google's investment in roadways is extractive in that it derives values from collecting data from its roads without re-investing in the public systems that maintain these infrastructures.[89] Like other Big Tech companies, including Amazon, Apple, and Meta, Google has avoided paying billions in taxes, reported by Safiya Noble as collectively $155.3 billion in taxes between 2010 and 2019.[90] While Google Maps' business model is based on making public infrastructures like roadways highly visible on its map, it conceals the public infrastructures and resources (such as roadways, water, and electricity) that these systems depend on and extract from.

In this sense, Google Maps' relation to publicness operates through what Shannon Mattern theorises as a practice of grafting.[91] Drawing from processes of plant propagation, Mattern demonstrates how grafting is both a creative and a manipulative process in the context of smart city interventions. In this sense, the scion and the stock of the plant continue to grow together as a form of regenerative growth. But at times the relationship is unidirectional, extracting rather than sharing. For Mattern, technology companies sell the imaginary of an easy grafting of "smartness" onto a city, a smartness that extracts data in the name of a city's smooth functioning.[92] More than simply a method of extraction, this type of grafting is "destabilising." Companies received federal tax subsidies, public grants, and publicly funded research to work.

Grafting's imbalance of resource distribution reflects Plantin, Lagoze, Edwards, and Sandvig's argument that platforms rise "when infrastructures splinter,"[93] (an adaptation of Graham and Marvin's of splintering urbanism).[94] Big Tech companies like Google benefit directly from the erosion of public services,[95] filling the vacuum of publicly accessible resources with free or low-cost tools[96] or selling systems to government that promise efficiency of service.[97] Google claims that "there's never been a more important time for us to be helpful" making its dominance of the public realm seem like more than simply a service. As such, the need for platforms like

Google Maps providing services seems inevitable when there is a lack of public resources. This inevitability conditions our relationship to these platforms and their profiting from public need seems deserved.

Google's claims to publicness often collide with its public grafting. This is, perhaps, most stark in Google's data centres, where Google houses and maintains its data for such platforms as Google Maps. The data centre is a powerful site, or "where the internet lives"[98] according to Google. And while powerful, it is also a secreted-away industrial node of the internet's functioning. The images Google does share make the data centre look clean, safe, and efficient, minimising its material presence in the name of being home to "the cloud."[99] Google's data centres are not only leveraged as a necessary (albeit often secreted and hidden) part of the infrastructure that undergirds Google's data-driven Ground Truth project (that information needs to be stored somewhere!) but the presence of data centres is often heralded as an economic boon for a local economy, providing jobs for people in so-called remote communities.[100]

Perhaps the most notable extraction of resources is from Google's data centres, requiring access to water for cooling. The location of data centres in places like South Carolina ensures they have access to tax breaks, space, and local water systems. James N Gilmore and Bailey Troutman have looked at Google's three-year process to gain permission to increase its water extraction from public resources to power and cool its data centre in South Carolina.[101] Drawing from local reporting, Gilmore and Troutman analyse the conflict of Google's 2016 proposal to withdraw 1.5 million gallons of water per day from an aquifer located near Charleston, South Carolina.

The aquifer is a crucial water resource for South Carolina.[102] Google wanted access to the aquifer to help cool and power one of its data centres. However, sustained reliance of the aquifer poses risks of water depletion in the area.[103] Yet, despite public pushback, Google presents itself as adding to communities rather than taking from them. For example, in 2016, Google claimed to have contributed $750 million to labour income in the United States, as well as 11,000 jobs. Gilmore and Troutman, in their analysis of local news in South Carolina, reveal how Google Maps uses the language of "neighbours" to present as members of the community, citing their invest-ments of $1.8 billion as evidence of being a "good neighbour."[104] While data centres extract resources from communities, Google is quick to position these same data centres as a public service, providing jobs and financial investment in place of simply taking from them. Google positions these natural (and in this case public) resources as sites of investment in the continuation and maintenance of its data fantasy.

Google's data centre in South Carolina is not an outlier. As Mél Hogan has shown how Facebook's data centre diverts 30% of the water from Georgia's Chattahoochee River for its server cooling.[105] Similarly, the city of Northlake in Illinois pumps its water to Microsoft's massive data centre in Chicago.[106] While Dallas-area data centres have looked to solutions like digging a 1200-foot-deep well and reusing the city's wastewater while it demands nearly 1.5 billion gallons of water per year for a new data centre, nearly a tenth of the water the county uses.[107] The stop-gap measures of building wells for drought become baked into the infrastructures that support data centres. Google not only relies on public water supplies it drains these supplies at the expense of public access.

The data centre is what Mél Hogan calls "the underbelly"—concealed and made to seem invisible.[108] The exterior presence of data centres is often nondescript, located out of the eye of public scrutiny. Data centres are often in suburban and rural areas where they have space and access to natural resources. Data centres are often located in cooler locations, such as Hamina, Finland,[109] where it is easier to cool the heat because of the lower temperatures of these surroundings. As the secreted away and heavily secured structure holding and storing Google's data, Google's data centres come to symbolise the absence of Google's public-facing engagement.

However, the data centre is also a material manifestation where users can locate Google, and they do so on Google Maps: in the absence of a formalised public engagement process, people have used Google Maps' review platform (described in more detail in Chapter 3) to "review" Google itself. The data centres on Google Maps in Lenoir, South Carolina or in Council Bluff, Iowa, for example, have been reviewed on Google much like a user might review another business or store. In other words, people "locate" Google in these data centres on Google Maps and use this as a space for public feedback. But how these data centres are found is an interesting part of attempts to contact Google. Google Maps positions itself as a public service but hides its material operations. Moreover, it provides no (or extremely limited) means for the public to respond to, critique, or adjust how they engage with Google.

Reading these reviews, we see ratings not specifically for the data centre, which is not open to the public, but ratings and reviews for Google itself. Indeed, many of these data centres have a low average rating based on the many one-star ratings (out of five) accompanied by angry reviews. Complaints include those of business owners who have received misplaced or mistaken bad reviews for similarly named businesses in a neighbouring city, or for a different business in the same building. The reviews of Google's data centres describe frustration with inaccurate reviews appearing on their business listing. These reviewers worry that such reviews threaten their businesses, since being on Google Maps and reviewable on Google Maps is essential to being seen by potential customers. Moreover, those leaving these reviews are annoyed that there is no way to contact Google to correct this.

There are also posts from people who are angry about their work and efforts of being a reviewer being erased. There are complaints about inconsistencies with the five-star rating system. There are issues with Google Maps itself, such as having trouble accessing their main account. Most significantly, the overarching issue is not being able to reach Google itself—a persistent echo of there being "no way to contact Google." One can register their complaint, but one cannot speak directly with "Google." While the angry reviewers are visible, Google is absent, trying to contact Google Maps to no avail.

These complaints—and the fact that they are delivered through the only means possible (reviews of Google's data centres)—reveal how Google manages conflict and disruption. As Sara Ahmed[110] argues, institutional complaint protocols contain critique in locations, and do not fundamentally disrupt the institution's ability to continue with "business as usual." Google is beyond reproach. These are closed matters, without public debate. Instead, complaints make the complainer visible to the system and those watching as a disruption of the peace and conviviality. Since the only recourse that Google leaves open is the map itself, the enclosure Google Maps imposes, returning to Plantin's term,[111] is totalising.

Google Maps uses public resources and positions itself as a public service but offers no avenue for public engagement, thus strategically dislocating itself from publicness. This makes its presence seem all the more inevitable: by carefully controlling and limiting public engagement, Google Maps makes it seem like there's no way for the public to question or change it. Google Maps' location awareness relies on existing public infrastructures to run, whilst extracting and exploiting public resources it claims to invest in. In the absence of public accountability, Google Maps frames itself as a public service. More than an eye from nowhere, Google is an administrative tool from nowhere that seems impossible to locate.

**Thank Goodness for Google**

What is taken for granted in the refrain, "thank goodness for Google?" More pointedly, what does Google take for granted in its self-positioning as a public resource? While Google leverages its map and mapping tools as beneficial for the public, these benefits are not evenly distributed. What does it mean to make a more "sustainable choice" about which route to take when the structures of racial capitalism and global logistics have foreclosed which areas are unsustainable for human life. It not only reifies unequal distributions of pollution but also contributes to this pollution.

Google Maps uses, extracts, and regulates public resources as a means of grafting onto pre-established notions of publicness. Google's infrastructural arrangements also organise publicness, through how they are deployed, what they extract, and how they evade. Claiming the status of a public resource becomes a means to use this status to extract from public sources whilst refusing accountability or public-facing conversations. Google's claims of location awareness, as such, are organised through logics of investment, from where to spend time, how to move through space, and how to "invest" in community via jobs.

## Notes

1 To animate this situation further, the directions did not include instructions about where to walk once I got out of Tottenham Court Road Station. And, indeed, I ended up walking in the wrong direction, away from the British Museum, and into Fitzrovia.
2 Google, "Our Approach to Search."
3 Graham and Dittus, *Geographies of Digital Exclusions.*
4 Google Maps quoted in Graham and Dittus, *Geographies of Digital Exclusion,* 10.
5 Reid, "How 15 Years of Mapping the World Makes Search Better."
6 Noble, *Algorithms of Oppression.*
7 Here I am drawing from two pieces, both by Scott McQuire: "Googling the City," in *Geomedia: Networked Cities and the Future of Public Space* (Cambridge, UK: Polity, 2016) and "Learning from Street View: Lessons in Urban Visuality," in *Visual and Multimodal Urban Sociology, Part A, Research in Urban Sociology, Vol. 18 A*, ed. Luc Pauwels (Bingley: Emerald Publishing Limited, 2023), 141–160, https://doi.org/10.1108/S1047-00422023000018A006.
8 McQuire, "One Map to Rule Them All," 150–165.
9 Berlant, "The Face of America and the State of Emergency," 177
10 Berlant, "The Face of America," 178.
11 Plantin, "Google Maps as Cartographic Infrastructure."
12 Pickles, *Ground Truth.*
13 Pickles, *Ground Truth*; Thatcher et al., "Revisiting Critical GIS."
14 Bennett, et al., "The Politics of Pixels," 729–752.
15 Parks and Kaplan, *Life in the Age of Drone Warfare*; Presner, Shepard, and Kawano, *HyperCities*; Kaplan, "Precision Targets: GPS and the Militarization of U.S. Consumer Identity; "Chow, *The Age of the World Target*; Virilio, *War and Cinema.*
16 Parks and Kaplan, *Life in the Age of Drone Warfare.*
17 Kaplan, "Precision Targets," 705–709.
18 Haraway, "Situated Knowledges," 582.
19 Bondi and Domosh, "Other Figures in Other Places," 199–213, Roberts and Schein, "Earth Shattering: Global Imagery and GIS," 171–195.
20 Kwan, "Feminist Visualization."
21 Kwan, "Feminist Visualization," Elwood, "Critical Issues in Critical GIS," Elwood and Leszczynski, "Feminist Digital Geographies," Tidy, "Gender Politics of 'Ground Truth,'" 99–114.
22 In this case, I am drawing from the work of Mirzoeff, *The Right to Look*; Haraway, "Situated Knowledges," 575–599; Denis Cosgrove, *Apollo's Eye*; Parks, "Satellite Views of Srebrenica," 585–611; Perkins and Dodge, "Satellite Imagery and the Spectacle of Secret Spaces," 546–560; Ash, Kitchin, and Leszczynski, "Digital Turn, Digital Geographies?" 25–43; Kwan, "Algorithmic Geographies," 274–282.
23 Wilken and Thomas, "Vertical Geomediation," 2531–2547; Lee, "A Political Economic Critique," 909–928; Farman, "Mapping the Digital Empire," 869–888.
24 Pickles, *Ground Truth*, 174.
25 Farman, "Mapping the Digital Empire," 869–888.
26 McQuire, "Learning from Street View: Lessons in Urban Visuality," 149. Also in McQuire's chapter "Google the City" in his book *Geomedia: Networked Cities and the Future of Public Space.*
27 Packer, *Mobility Without Mayhem.*
28 Google, "Bringing Your Map to Life."
29 Russell, "How Google Street View Mapped the World."
30 Russell, "How Google Street View Mapped the World."
31 Luque-Ayala and Neves Maia, "Digital Territories: Google Maps as a Political Technique," 449–467. This is further discussed in Chapter 5.

32 Cheung, "Mapping Stories with a New Street View Trekker."
33 Madrigal, "How Google Builds Its Maps and What It Means for the Future of Everything."
34 Madrigal, "How Google Builds Its Maps."
35 Google, "Project Ground Truth."
36 Miller, "The Huge, Unseen Operation Behind the Accuracy of Google Maps."
37 Madrigal, "How Google Builds Its Maps."
38 Mulvin, *Proxies*, 47.
39 Gillespie, *Custodians of the Internet*; Roberts, "Social Media's Silent Filter"; Roberts, *Behind the Screen*.
40 Introna and Nissenbaum, "Shaping the Web," 169–185; Noble, *Algorithms of Oppression;* Vaidhyanathan, *The Googlization of Everything*.
41 Justie, "Little History of CAPTCHA," 30–47.
42 Graham and Dittus, *Geographies of Digital Exclusion*.
43 Graham and Dittus, *Geographies of Digital Exclusion*.
44 Plantin, "Google Maps as Cartographic Infrastructure."
45 Plantin, "Google Maps as Cartographic Infrastructure."
46 Cowen, *The Deadly Life of Logistics*.
47 Hockenberry, Starosielski, and Zieger, "Introduction: The Logistics of Media," *Assembly Codes*, 1.
48 Google Earth Outreach, "Air Quality."
49 Dicker, "Navigate More Sustainably with Maps."
50 Phillips, "Maps Is Getting More Immersive and Sustainable."
51 Phillips, "Maps Is Getting More Immersive and Sustainable."
52 Phillips, "Maps Look and Feel More Like the Real World."
53 Aclima, "Built for Good."
54 "Labs: Air Quality," Google.
55 Google, "Air Quality."
56 Google, "Air Quality."
57 Google, "Air Quality."
58 Liboiron, *Pollution Is Colonialism*.
59 Gilliard, "From Redlining to Digital Redlining."
60 Gilliard and Culik, "Digital Redlining, Access, and Privacy."
61 Squires, "Racial Profiling, Insurance Style," 391–410; Harris, *Little White Houses*; Light, "Discriminating Appraisals," 485–522.
62 Gilliard, "Pedagogy and the Logic of Platforms," 64.
63 Fiske, "Surveilling the City," 71.
64 Othering and Belonging Institute, "Segregated Cities in 2020"; Menendian, Gambhir, and Gailes, "Racial Residential Segregation."
65 This is discussed in more detail in Chapter 5. Also see Robert K Nelson, LaDale Winling, Richard Marciano, and N.D.B. Connolly, "Mapping Inequality" a partnership between the University of Richmond, Virginia Tech, University of Maryland, and Johns Hopkins University, https://dsl.richmond.edu/panorama/redlining/#loc=4/41.212/-112.72; *American Panorama*, ed. Robert K. Nelson and Edward L. Ayers, https://dsl.richmond.edu/panorama/redlining/.
66 Cortright, "Where Does Houston Rank"; Kinder Institute, *Houston Disparity Atlas*; Kinder Institute, *Houston Region Diversity Report*.
67 Hays, Landeiro, and Rongerude, *Neighborhood Knowledge for Change*.
68 Hays, Landeiro, and Rongerude, *Neighborhood Knowledge for Change*; Palaniappan, "Ditching Diesel," 31–34; Gonzalez et al., "Research and Policy Advocacy to Reduce Diesel Exposure," S166–75.
69 Hays et al., "Neighborhood Knowledge for Change."
70 Palaniappan, "Ditching Diesel," 31–34.

71  Gilliard, "Pedagogy and the Logic of Platforms"; Noble, *Algorithms of Oppression;* Safransky, "Geographies of Algorithmic Violence," 200–218.
72  Gilliard, "Pedagogy and the Logic of Platforms."
73  McMillan Cottom, "Where Platform Capitalism and Racial Capitalism Meet."
74  Safransky, "Geographies of Algorithmic Violence."
75  Elwood, "Digital Geographies, Feminist Relationality, Black and Queer Code Studies."
76  Liboiron, *Pollution Is Colonialism.*
77  Liboiron, *Pollution Is Colonialism*; Tuck & McKenzie, *Place in Research.*
78  Belfiore and Bennett, "Determinants of Impact," 225–275; Gudeman, "The New Captains of Information," 1–3.
79  Oman, *Understanding Well-Being Data.*
80  Hicks, "Risky (Information) Business," 1157.
81  Hoffmann, "Terms of Inclusion."
82  Garzon et al., *Community-Based Climate Adaptation Planning.*
83  The Oakland Climate Action Coalition, "Oakland Climate Action Coalition"; Gonzalez, *Community-Driven Climate Resilience Planning.*
84  Loukissas, *All Data Are Local.*
85  Safransky, "Geographies of Algorithmic Violence."
86  Safransky, "Geographies of Algorithmic Violence," 201.
87  L'Oréal Groupe, "New Strategic Partnership with Israeli Climate Tech Company"; Hathaway, "L'Oréal Enters Strategic Partnership"; Caldwell, "L'Oréal Signs Strategic Partnership with Breezometer."
88  Google Could, "BreezoMeter: Delivering global environmental information with Google Cloud."
89  Noble, "The Loss of Public Goods"; Graham and Dittus, *Geographies of Inequality.*
90  Noble, "The Loss of Public Goods."
91  Mattern, *A City Is Not a Computer.*
92  Martin, "Shipping Container Mobilities," 1021–1036; Kitchin, "The Real-Time City?" 1–14.
93  Plantin et al., "Infrastructure Studies Meet Platform Studies," 299.
94  Graham and Marvin, *Splintering Urbanism.*
95  Noble, "The Loss of Public Goods."
96  Mattern, "Maintenance and Care."
97  Eubanks, *Digital Dead End.*
98  Fischer, "Unseen World of Data Centres"; Holt and Vonderau, "Where the Internet Lives," 71–93.
99  Holt and Vonderau, "Where the Internet Lives," 71–93; Carruth, "The Digital Cloud," 339–364.
100 Gilmore and Troutman, "Articulating Infrastructure to Water," 916–931; Wong, "A Sustainable Solution"; Wong," Data Centers Change the World around Them"; Junius, "New Data on Data Centers."
101 Gilmore and Troutman, "Articulating Infrastructure to Water"; Larkin, "Politics and Poetics of Infrastructure," 327–343; Martin, "Controlling Flow," 147–159; Packer, "What Is an Archive?" 88–104; Parker, "Containerisation," 1–20; Siegert, "Map Is the Territory," 13–16; Young, "Cultural Techniques and Logistical Media."
102 Kingsbury, *Groundwater Quality.*
103 Gilmore and Troutman, "Articulating Infrastructure to Water," 73.
104 Gilmore and Troutman, "Articulating Infrastructure to Water."
105 Hogan, "Facebook Data Storage Centers," 3–18; Hogan, "Data Flows and Water Woes: The Utah Data Center."
106 Voices of the Industry, "Chicago Is a Geostrategic Destination."
107 Sattiraju, "The Secret Cost of Google's Data Centers."

108 Hogan, "Facebook Data Storage Centers," 3–18; Jacobson and Hogan, "Retrofitted Data Centres," 78–94.
109 Google Data Center, "Hamina Finland: From Paper Mill to Data Center."
110 Ahmed, *Complaint!*
111 Plantin, "Google Maps as Cartographic Infrastructure," 492.

# 3    Geographies of Self-Sufficiency

## Exploration | Experience

I stood on the corner of 15th Street and 10th Avenue in New York City, outside of Chelsea Market, at the food hall the size of an entire city block housed in a former Nabisco cookie factory. Setting the boundary for the Meat Packing district, the docks, and Hell's Kitchen, Chelsea Market is now a place of food stalls and clothing stores, across the street from Google's gold high rise—a reminder of Alphabet's ownership of the market.[1] I was looking for directions to Washington Square Park in Greenwich Village, in lower Manhattan, the famed gathering space of poets, folksingers, and chess enthusiasts. Together, Chelsea Market and Washington Square Park are symbolic pillars of New York's gentrification and change—sites of expensive housing and expansive commercial endeavours that have pushed out and flattened many of the racialised, queer, and creative cultures that animated these neighbourhoods, from The Stroll to the Club Kids. Being in and moving through these locations, I am reminded of the explicit, implicit, and enduring claim to space that animates modes of occupation and mobility.

Moving along 15th Street, I approached someone walking by and asked for directions. *It's over there*, he announced while pointing into the distance, creating a line with his index finger that pierced through the office towers, apartment blocks, and storefronts standing between us and the destination. He routed me through a list of streets, instructing me to head down 9th Ave, along 14th Street, and then to turn right at 5th. I couldn't keep track of his sequence of navigational commands, nor the tacitly held shorthand of New York City where "Street" signals an east/west direction and "Ave" a north/south orientation. And which one was 5th? I asked him to repeat his directions and (if he'd be so kind) *draw* them out for me with the paper and pen I retrieved from my bag. Ignoring my request for a drawing, he took out his phone and opened Google Maps. As the map loaded, he asked me where I was from. *I'm from Toronto*, I replied and then after a short but awkward pause I hurriedly added, *I can't use my data here.*[2] He responded, *Oh, that makes sense. I always figure something must be wrong with people's phones when they are asking for directions.*[3]

DOI: 10.4324/9781003251569-3

In that beat, I realised asking for directions had located *me* as out of sync with his understanding of wayfinding norms. What was wrong with my technology? I should be able to open Google Maps on my phone (like he was doing, in that moment) and seamlessly move through the city. Right? I was insufficiently self-sufficient.

Variations of this encounter occurred throughout my project of asking for directions in New York, London, Amsterdam, and Toronto. The people I stopped often asked questions like, what's wrong with your phone? Did you check Google Maps? Or why don't you look it up yourself? On the surface, these questions reflect an ordinariness of using Google Maps to find one's way.[4] But digging a bit deeper, the assumptions baked into these questions—that these directions should be simply at hand via the digital map on my mobile phone—elucidate the types of access to location and location awareness imagined in these moments of navigation. Such access to digital maps also illustrates the proficiencies projected onto the map itself. We don't just consume that expertise as users of the map; that expertise is made part of our own self-sufficient mobilities by not only granting purchase to Google's spatial knowledge but also on space itself. In short, the question "what's wrong with your phone" presumes that this possession of expertise is transferable through the tools we use, reflective of Google's claims of producing "maps for everyone, everywhere."[5]

Scrolling through the Google Maps application or website, it is difficult to avoid Google Maps' hallmark promises of exploration and access to experience. Calls to "explore and experience the world," "discover new experiences," "find places for you," or "have the confidence to explore," command imaginaries of unrestricted entitlement and complete control over one's personal domain—desirable mobilities available to everyone via the platform. From its software to its interface to its marketing, Google Maps' imperatives of *explore*, *experience,* and *access* structure engagement with the Google Maps platform and the world it mediates.

Explore, experience, and access organise space according to the terms of technologically mediated self-reliance where one can access anywhere "however you like" *without the need to ask anyone for help along the way*.[6] These imperatives organise space according to the right to be anywhere and assert one's presence while also being the expert of that place. Google Maps grants entrée to space via its command of location awareness, readily available through one's mobile phone. But while all space seems claimable, this "anywhere" is not available to everyone. Rather than a flexible and free relation to space, the fantasy of a self-sufficient user models an entitlement to space, a brand of entitlement that foregrounds colonial prerogatives of entry, ownership, and occupation.[7]

**Exploration: Make the World Your Own**

In his 2017 Google I/O presentation titled, *Making the World Your Own with Google Maps APIs*, Google Map Product Manager Joël Kalmanowicz invites the audience to (fittingly) "make the world your own, however, works best for you."[8] Throughout his talk, Kalmanowicz illustrates the elastic scaling of the Google Maps promise—from the total-world omniscience to the immediate awareness of one's exact surroundings. All spatial information can be made available in an instant, and the world can be made *your own*. Kalmanowicz includes looking up sushi restaurants near the Google I/O conference facility in Mountain View, California, or looking up "restaurants near a hotel you just booked." Trading in the language of ubiquity and availability paired with consumption and leisure, Google Maps produces orientations of everyday self-assured navigation because "if it's on Google Maps, you can find it." The extent that everything is available including the immediacy of experiences grafts ordinary navigation practices like finding one's way onto a boundless sense of entitlement where *everyone* can search for and go *anywhere* and do *anything*.

   *Explore* is a command nested into the Google Maps interface with invitations to *discover* and *explore*, organising how to *search* for geographic information (See Figure 3.1). The main mapping page, presented when

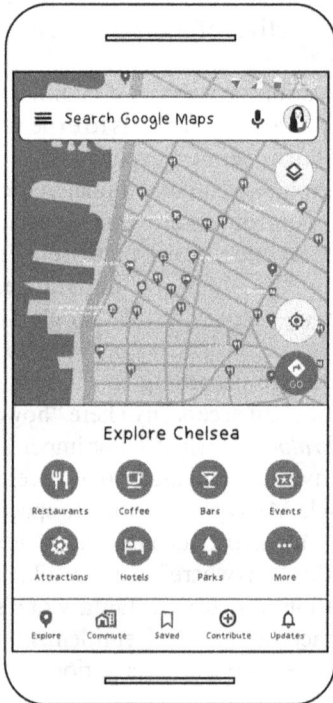

*Figure 3.1* Illustration of Google Maps' Explore Page with the invitation to "Explore Chelsea" with clickable icons for Restaurants, Coffee, Bars, Events, Attractions, Hotels, Parks, and More. Illustration by Colin Medley.

opening Google Maps, is the Explore Page, promoted on the Google Maps About Page as a means to "discover new experiences across the world and around the corner."⁹ Opening the application on a mobile phone, I am greeted with its signature vector map—with its standardised colour palate applied to all places that Google maps. I see the blue orb of location awareness pulsing away on the map of my neighbourhood. The blue orb's heart is beating in anticipation of somewhere else, directed by the icons, tabs, links, and symbols that frame and annotate this map's template. The page is annotated by directives to explore, discover new, search here. At the bottom of the screen, I am given a menu to access five pages: "Explore"—the landing page—"Go" (formerly Commute), "Saved" (formally For You), Contribute, and Update. The Explore Tab is marketed to "find places to eat and things to do around you or when you travel." I can scroll down to see the "latest" in my neighbourhood by way of recommended businesses such as bars and restaurants. The interface foregrounds the ease and convenience of exploration as mediated through Google Maps, and the geographies of entitlement assumed through this process.

Google's brand of *exploration* carries with it assumptions of unbridled access for the self-sufficient user. A type of access categorised by "restaurants," "coffee," and "parks" pinned to the map, like that seen in the invitation to Explore Chelsea. But in these projected geographies of self-assured navigation, who is the prototype that fulfils the fantasy of seamless mobility and unrestricted claims to space that can follow the routes marked by Google? While exploration may seem conventional to moving through the world—part of seeing and experiencing new places and things—we nevertheless must be wary of the types of claims Google Maps makes through this technology of self-sufficient "exploration." The invitation to explore expands the domain of the self-sufficient user beyond the everyday and routine and organises the entire world according to the terms that space can claim, everywhere can be known, and anywhere can become "your world." Ultimately, Google's imaginary of untethered exploration—the experience and access it engenders—carries a heavy weight.

**Why *We* Map the World**

Nestled at the bottom of Google Maps' "About Page"—found if scrolling past the promises to "explore your world" and "navigate the world around you"—is a video titled, "Why we map the world."¹⁰ This video follows Google Engineering VP Luiz André Barroso as he walks through the Google Campus, and narrates a "history" of mapping. The video's subheading, "Map making is an ancient human endeavor, and one that those of us working on Google Maps are honored to continue to pursue,"¹¹ sets the course of Barroso's cartographic history. Barroso takes the viewer through what he describes as a 2600-year mapping journey that culminates in digital mapping, or, specifically, Google Maps. Barroso's main claim is that mapping

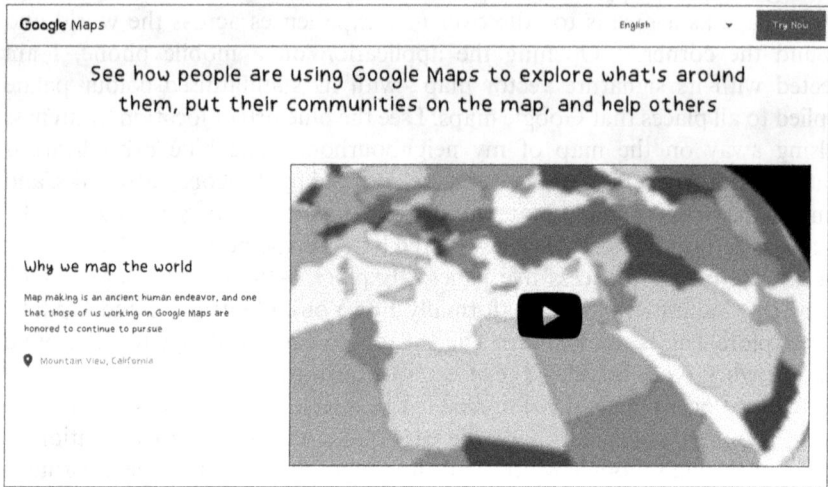

*Figure 3.2* Illustration of Google Maps' About Page, where one can see the video: "Why we map the world." The video description reads: "Map making is an ancient human endeavor, and one that those of us working on Google Maps are honored to continue to pursue." The header for the section reads: "See how people are using Google Maps to explore what's around them, put their communities on the map, and help others."

is "one of humankind's driving forces" and that even after millennia of mapping projects, "the job isn't done." The video's premise is two-fold: (1) mapping is not simply a representation but an enduring means to access truth about the world; (2) Google Maps is the natural inheritor of this time-honoured universal mapping project. The rationale for why Google maps the world—and its understanding of how those activities fit into the history of mapping—locates exploration as a form of progress (Figure 3.2).

Barroso's abridged timeline of mapping technologies starts in "600 BC" from what he terms the "first map we know of." This first map, often referred to as the Babylonian Map of the World, is a palm-sized clay tablet map (laying the visual groundwork for the smartphone continuum). From the tablet, Barroso narrates a whistle-stop tour of navigational technologies, including the invention of latitude and longitude, and the development of the magnetic needle to mark *North*. Then he moves through what he calls "a burst of mapping activity in the 15th century" one that "expanded the boundaries of the documented world."[12] Barroso describes the development of the marine chronometer as "opening up distant horizons," which he claims, "further accelerated the Age of Discovery."

He then skips to the first aerial imagery from hot air balloons in the mid-1800s then over to the 20th century which he situates as moving through "exponential progress" based on the first computer-generated map of the 1950s, satellite mapping in the 1970s, and GPS in the 1980s. Funnelled through this narrative, Barroso historicises these spatial technologies as leading towards a natural course of advancement, culminating in 2005 where Google "joins the story" with the first local search index to catalogue locations of businesses (besides the Yellow Pages, I suppose). Barroso concludes with an overview of the technological contributions of Google Maps' project Ground Truth[13] to Google's ambitious project of producing what he describes as a "useful and meaningful map" of the whole world.

Barroso's narrative tracks a linear progression of mapping technologies culminating in the digital tools of today. The video consolidates a multitude of mapping technologies into a single trajectory towards Google Maps and its mission to help millions of people every day to explore the world. The video positions mapping and global exploration as part of an inevitable desire to know and navigate the world—a depoliticisation of mapping made all the more glaring in the absence of opposition. These technologies do not simply document the world but measure and standardise an understanding of it.

These discursive manoeuvres of "expansion," "opening up," and "Age of Discovery" conceal the violence of global mapping projects. This language tempers what Anna Lauren Hoffman terms the "discursive violence" of data-driven systems like mapping, and also normalise the ontologies driving these systems of knowing the world. Barroso mobilises the utterance of "the age of discovery" to suggest a shared history of cartography, erasing the driving force of colonialism at the helm. Eve Tuck and K. Wayne Yang, Leanne Betasamosake Simpson, and Linda Tuhwai Smith, among other Indigenous and anti-colonial scholars, have repeatedly illustrated that imperial "discovery" is not a singular brutal act but a structural violence and an ongoing process.[14] As Tuck and Yang articulate, colonial violence is a force that extends beyond the arrival of the so-called explorers, "but is reasserted each day of occupation."[15] Barroso calls this time a period of "rapid progress" but fails to connect the precipitous interest in the development of mapping technologies with the desire of European states to outpace one another as they lay claim to the rest of the world. This language illustrates how embedded claiming space is in this linear narrative of mapping.

While Indigenous people have long challenged the colonial processes of mapping and classification of the world[16] (recalling Simpson's concept of Nishnaabeg grounded normativity) Google Maps continues to organise space according to these terms of the brave explorer. Barroso euphemistically labels the expansion of global trade that marked genocide, slavery, and extraction as a "burst of activity" and in so doing erases the presence of resistance and Indigenous existence. The thrust of mapping is unquestioned and redeploys the colonial "discovery" narrative as uncontested truth. Progress and innovation are untethered from their intention to control, dominate, and possess. Google's claims to this lineage of mapping practices situate the prototypical self-sufficient explorer as one who both "opens up distant horizons" and moves seamlessly through an imagined elsewhere. The exceptional explorer is a long-exercised colonialist trope baked into the language of Christopher Columbus's voyage to "the New World" or the Lewis and Clark "expedition" to the "American Frontier."[17]

The heroic explorer prototype and the spatial occupations they engender have long been refused and rejected.[18] Indigenous-led protests of civic holidays like the USA's Columbus Day, demonstrate that "the Age of Discovery" did not simply happen in the past but continues to happen.[19] While refusal of colonialism takes many forms, or what Leanne Betasamosake Simpson theorises as a grounded ethics continually generated in relationships with land and other beings,[20] one symbolically evident rejection of the explorer prototype is the toppling of several "explorer statues" including the James Cook statue in Victoria, British Columbia,[21] the Christopher Columbus statue in St. Paul Minnesota in 2020,[22] Sebastián de Belalcáza in Popayán, and Columbus in Barranquillam Colombia.[23] The ongoing actions as resistance show that explorers are not imaginary figures and their actions of "exploring" are not benign metaphors.[24] These are acts of violence with continuing impacts that, as Eve Tuck and Marcia McKenzie argue, "continue to dispossess Indigenous peoples and Black peoples, promoting white supremacy."[25] And while the statues might topple, the video stays up on Google Maps' website to promote the idea of its universal world map as the product of a shared history of exploration. The explorer is not a neutral character, yet it remains a central figure of Google Maps' territory.

Exploration propels an imaginary of boundless access and limitless discovery made possible through settler colonial assumptions about spatial entitlement. Along these lines, Google's motif of *exploration* parallels Jas Rault's analysis of *transparency* as a communication technology that leverages promises of access while at the same time obscuring the power of entrenched colonial administration that controls transparency's terms.[26] As Rault argues, transparency promises full access to truth and accountability, but the lines by which this truth is constructed and understood remain obscured. Rault cites the example of public consultations as one of the techniques of transparency, where carefully framed conversations with the public become an ancillary form of listening without necessarily having any intent to act or meaningfully recognise the power structures at play even in acts of "listening." In one example, Rault points to Canada where settler government consultation with Indigenous communities normalises and validates settler colonial presence. Rault describes such consultations as "nonreciprocal transparency: a performance and aesthetic of settler colonial openness."[27] Transparency provides access to an already-closed conversation and so enforces asymmetrical relations, promising *access to* or *purchase on* information but actually holding the power to control the terms of that access.

Rault's framework re-orients Google's conceit of exploration—reflected through Barroso's bravado of opening up the world and providing access to spatial information—as a similar imperialist tool of possession and entitlement. Google's brand of exploration is in fact a technology in itself: it is something Google uses to rationalise and justify the Google project of a complete map and the processes needed to get there. Barroso's narrative reveals how exploration has long been the justification behind mapping, obscuring the narrative of control. Exploration as spatial technology centres settler world-making and emplacement, making claims on what is new, what is unknown, and what is undiscovered. And further still, it is the intrepid explorer set to know the world—a self-sufficiency of complete knowing. In both cases, concepts of transparency and of exploration are leveraged as technologies that progress and extend access assuming the right designs and affordances are at play.

Barroso's video not only promotes mapping as what Sandy Grande calls a "colonial consciousness"[28] but also promotes the exlporatory function of the map as the "settler common sense." The map's role in laying claim to place relates to what Mary Pratt describes as "European planetary consciousness based on the dual impulses towards interior exploration and constructions of global-scale meanings."[29] This trajectory of mapping sets the scene for how Google Maps frames exploration as a valiant discovery rather than a method of dispossession, making not only mapping but spatial domination appear natural and inevitable rather than contested and refused.[30] Google Maps' brand of exploration maintains and normalises what Tuck and McKenzie term the "structured antagonisms"[31] of assumed blanket entitlement to all land.

By situating Google Maps in a linear sequencing of mapping innovation, driven by a voracity for "knowing the world," Barroso presumes an inevitability of the drive for imperialist control and settler ascendancy and in doing so, makes light of the enduring violence enacted in the name of exploration, claim, and discovery.[32] But more insidious is the scaling of mapping and knowing the entire world to the scope of everyday use. Getting directions to work, travelling on vacation, to visiting new places—these processes of making it your own—are entangled in claiming the world as yours. These are the assumptions that, to use the words of Dylan Mulvin, "stick to and travel with" these prototypes.[33]

### Exploration and "What the World Has to Offer"

More than the general promise of exploration, Google Maps promotes itself as a means to *be* the explorer—the person, ostensibly "anyone," who can go "anywhere." Google Maps Vice President Jen Fitzpatrick pledges that Google Maps is "making it easy for you to explore and experience what the world has to offer." In her *I/O* Keynote, Fitzpatrick takes to the stage accompanied by an introductory video celebrating the "power" of Google Maps. The video runs through montages of landscapes accompanied by pensive piano arrangements and an overlayed text which reads: "They guide us and connect us. And Google took us where they had never been before. Everywhere. And to everyone."[34] The video sets the scene of access to these diverse landscapes, from remote mountain top to a busy city. It crafts a narrative of "elsewhere" that the fantasy of a universal "everyone" can connect to.

Following the introductory video, Fitzpatrick delivers similar promises of universal access, unbridled exploration, and *travel confidence* with statements such as: "maps were built to assist everyone wherever they are in the world" and "we've given more than a billion people the ability to travel the world with the confidence that they won't get lost along the way."[35] The framing of confidence set against the unknown world begins to shape who this "everyone" is—the intrepid traveller who can voyage into the boundless anywhere. Similar promises are echoed at other Google events including the 2016 *Horizon Cloud*[36] meeting—advertised as an "enterprise consumer event"[37]—Fitzpatrick made the same promise that Google helps "everyone navigate, explore, and decide with confidence."[38] She adds that Google Maps is a "trusted guide to the real world that they can bring anywhere." By Fitzpatrick's logic, the map is an ingress to the phantasm of total world proficiency; but this "power of the map" fails to contend with the power differential framed by and produced through Google Maps as an instrument of exploration.

Exploration as an analogy for travel operates as both a spatial practice and a personal identity. Following Sara Ahmed, the fantasy of inclusion—marked by the hollow promise of "everyone" and "anywhere"—is a technique of exclusion.[39] There are a host of barriers that prohibit such boundless mobility. These barriers are geopolitical and biopolitical: passports, visas, personal records, and access to leisure time, travel insurance, and wealth. They are the controlled privileges to take planes, leave support networks for long or even short amounts of time, to be without immediate care responsibilities, access to wealth for leisure, a job one can take a break from, and the list goes on. And while people make it work despite all these imposed barriers, global mobility requires more than the confidence of direction or assurance that one will not get lost. In short, Google's prevailing promises of unbridled access to explore anywhere are tethered to a colonial consciousness of discovery made to seem inclusive and flexible.[40]

From package deals for climbing Kilimanjaro (thanks to the help of the unnamed guides), to the dispossession of resort tourism—travel has long been a means of staking claim on space. What these fantasies leave out are the border regimes that make travelling the world possible for some and impossible and dangerous for others. Harsha Walia's *Border and Rule* is an important corrective to Google's organisation of space. Walia describes relations to space as a "system of global apartheid, where displaced people are racially ordered and segregated as superfluous, capitalist techno-solutionism presented to solve what it has created by trading in a market of dispossessions, imperial states spuriously claim to care about refugees without sullying their own heavily guarded sovereignty, and elite humanitarianism is positioned as more pragmatic than meaningful justice."[41]

Global mobility and crossing territory are carefully managed and administered privileges. The borders themselves, built and conceptually rendered, generate illegality, and define trespassing. Further, as Walia demonstrates, border governance is exercised through strategies of exclusion, territorial diffusion, commodified inclusion, and discursive control. For many, the realities of crossing borders are a dangerous passage and for many the borders they live within are imposed. What is self-sufficiency within this reality except the mark of unacknowledged privilege and entitlement in a context in which "immigration and citizenship have been specifically weaponised to further the genocidal elimination of Indigenous political and social formations."[42] Google Maps reveals the stark contrast between individual freedoms of movement that further loots the geopolitical trenches that already exist. These discrepancies reflect what Doreen Massey terms the power geometry of space, which as she argues, "concerns not merely the issue of who moves and who doesn't, although that is an important element of it; it is also about power in relation to the flows and the movement."[43]

Google's declarations of seamless exploration collapse ordinary acts of getting around one's hometown with fantasies of discovering an imagined elsewhere, thereby tethering the pragmatics of everyday navigation to the assumption that all space (even illusory space) is available to claim. Fran Tonkiss describes similar techniques of unbridled passage through the city (here thinking of the wandering flaneur)[44] as the projection of "a particular masculine subjectivity onto urban sidewalks."[45] This delimits what can go past unseen, who can move through in the comfort of anonymity. But the repeated assumptions of full command to space, both familiar and unfamiliar, carry with it the norming logics of privileged access.

Simultaneous to the promise of exploration is the ongoing negotiation of space based on what Tonkiss terms are personal and political "cognitive maps of safety and danger" organised around calculations and geographies of risk.[46] For example, writer Garnette Cadogan describes his experience of "walking while Black" through New York City, as "a pantomime undertaken to avoid the choreography of criminality."[47] Cadogan illustrates his own mapping of space when he describes, "Walking as a Black man has made me feel simultaneously more removed from the city, in my awareness that I am perceived as suspect, and more closely connected to it, in the full attentiveness demanded by my vigilance. It has made me walk more purposefully in the city, becoming part of its flow, rather than observing, standing apart."[48] John Fiske, speaking further to these discriminatory spatial codes says: "street behaviours of white men (standing still and talking, using a cellular phone, passing an unseen object from one to another) may be coded as normal and thus granted no attention, whereas the same activity performed by Black men will be coded as lying on or beyond the boundary of the normal, and thus subject to disciplinary action."[49]

Robin Maynard's research on policing in Canada illustrates how institutional anti-Black racism blockades access to public space. For example, Maynard reports that in Halifax, Nova Scotia, Black people were three times more likely than white pepople to be stopped by the RCMP for street checks between 2006 and 2016, statistics consistent for other Canadian cities like

Kingston and Montreal.[50] Maynard also points to Toronto, where Black men are *carded*, a stop-and-frisk police tactic of "random" ID checks, at a rate of 3.4 times higher than that of white men in the city. The frequency of these checks is even higher in predominately white neighbourhoods. These figures reveal that being in space is not the same as accessing space. As Maynard writes, "while the ability to walk freely in public space is something that is taken for granted by most white Canadians, the same cannot be said for people of African descent."[51] This presumed neutrality of space, where whiteness is the default, reflects a racist mediation of mobilities that Sarah Sharma and Armond Towns have elsewhere argued "*assumes* a material relationship to autonomous movement while simultaneously controlling the movement of Others."[52]

Catherine McKittrick argues that Black geographies, ways of being in the world, resist and refuse these hegemonic organisations of space.[53] The activism of Black Lives Matter saw a reoccupation of the streets calling out racist politics and drawing attention to the spatialisation of racism. Alongside the reoccupation, there were important practices of resistance and care that are not meant to be visible to hegemonic structures—what Rianka Singh terms "resistance in a minor key." For Singh this resistance involves "the tactics employed by those for whom amplification is not the goal. It is a turn toward careful quietness. Sometimes this means refusing normative ways of using technologies, other times this means recognising that in this age of amplification that has led to the celebration of platforms as tools for making voices heard, visibility becomes antithetical to the survival and care of particular communities."[54] Visibilty is antithetical to survival in multiple senses here, as Tonia Sutherland reminds us, since big tech companies profit from both Black death and resistance as platforms that mediate witnessing and conversation.[55] Google's equating exploraton with spatial mobility (it's for everyone! To go anywhere!) erroneously consolidates seamless access to space with entitlement and individual power. Google's call to make the world your own fails to contend with "concealment, marginalization, and boundaries"[56] as spatial processes that in themselves have the power to rupture Google's imaginaries of claiming space.

Returning to Kalmanowicz's call to "make the world your own," a precise realisation of a self-sufficient prototype is Simon Weckert's artwork Google Maps Hack.[57] In his artwork from 2020, before much of the world went into lockdown, Weckert generated a "virtual traffic jam" through the manipulation of Google's traffic data. By walking along the streets of Berlin, Germany with 99 mobile phones open to Google Maps (all carried in a red wagon he held in tow), Weckert "tricked" the Google algorithms into thinking that 99 cars were jamming the same road he walked along freely. The accompanying video shows how Google Maps marked the streets in red to signify heavily congested traffic where Weckert walked, transmitted to drivers using real-time feedback of Weckert's imposed traffic conditions. As such, drivers diverted their path away from the pile of mobile phones being dragged through the streets. Weckert's work is revealing of the fantasy that underlies the self-sufficient explorer position—empty space. Weckert's project fulfills the fantasy of the Google Maps prototype: untethered access to the entire colonial city of Berlin, claimed at the hands of the tech-savvy, white man. Moreover, Weckert's project characterises the sharp contrast between freedom of movement and accessibility as he is able to freely move through the city at the expense of everyone else's flow. Where the rest of the world are antagonists in one's orchestrations of movement, he is the fully realised self-sufficient prototype at the helm of a mapping perpetuated by claims to space articulated through systems of whiteness.

What are the limits to unbridled exploration in a world encoded with systemic anti-Indigenous and anti-Black racism—where mobility can be tied to whether or not one appears criminal, threatening, or out of place?[58] This coding of space privileges what Safiya Noble and Sarah Roberts identify as the "imaginary of whiteness and unbridled exploration and intrusion,"[59] not only reinforcing the colonial structures but making them conditions of location awareness. Therefore, the self-sufficient explorer is a prototyping of only the most privileged called upon to take up the mantle of explorer when enlisted through Google Maps' model. This is not to say that the affordances of Google Maps cannot be adopted to fulfil needs. It is a tool that can be used in personalised ways to move through space differently in ways that may not even be legible to Google Maps itself (one can hope). People can picture a route before going, find out exactly where something is so they can walk with confidence, see if a location is accessible, see if there is subway access, make decisions about their personal safety, which may not be a formalised feature of the map. However, location awareness as a vehicle for self-sufficiency defined through exploration reduces what spatial access really means or really could and should be. Instead, Google Maps' brand of exploration grants access to many of the people who already have the all-access pass. Google's self-sufficient explorer, modelled through the commands of the Google Maps interface, presumes an extrapolation of the prototype.

### Experience: Accessing Time

Like the promise of exploration, the promise of "an experience" and "experiencing" what the world has to offer become the means through which to rationalise and justify the Google Maps project. Appeals like "We want to make it easy for you to explore and experience what the world has to offer" and "Discover new places & experiences" demonstrate how Google frames experience as the result of exploration, in language of positivity and conviviality. While exploration is the promise of access, experience is the currency of self-sufficiency, a purchase on an expertise of space that Google Maps deploys experience as a promise of enhanced self-sufficiency—enabling the user to experience the best food, the fastest route, or the newest bar through the platform. And while these promises are nice in the moment, they also overdetermine the types of relationships with space and the types of spaces worthy of mapping.

The self-sufficient connoisseur is oriented by "the best," "the new," and "your favourites."[60] In February 2020, celebrating its 15-year anniversary, Google Maps produced a 30-second video titled *Google Maps, There's More to Explore*.[61] The short video is a montage of different places to "experience in the city," includinge coffee shops, art museums, "the *best* tacos in town," or your "*new favourite* park"[62] (emphasis my own). The video charges the viewer to "beat the rush" and to "pick a table at the *tastiest* pizza place," collapsing work and leisure by pairing commuting to the office with enjoying a nice meal. The video concludes by summarising the scales of experience made possible through Google Maps: "From getting you around the corner to helping you discover your world, there's more to explore with Google Maps."[63] Experience is both everyday and extraordinary, framing routines of urban living as something that can be bested— reflective of what Doreen Massey calls the "conquering of time"[64] or the claims one makes on the times of others to facilitate this fantasy of a so-called sped up world.[65] As such, space is temporally controlled through the self-assured acts of getting ahead when one *beats the rush* or even reserves a table at the *tastiest pizza place.*

Seemingly clear-cut qualifiers like "the best" place or "the right" way demonstrate how experience is mobilised as a method of taming the environment. Google Maps' articulation of experience organises space according to absolute terms geared around self-optimisation and individualised productivity, asserting that there is a correct way of encountering the world or moving through it—thereby also implying an incorrect way. I am reminded of when I was in downtown Toronto by City Hall, looking for directions to St. Lawrence Market.[66] The person I asked for directions used their phone. They explained as they looked up the directions on Google Maps that they wanted to find the "exact" location. During a follow-up interview, they stressed that they "always" use Google Maps and excitedly described it as their "favourite app." Google Maps was part of their everyday routine, used even when they were familiar with the area. When I asked why, they recalled how they used it to find "the best way" to get home from work, stating that they became upset when they did not take the best route: "I guess I just want to make sure that if I know there are multiple routes, let's say home from work, and I typically leave around the same time every day but sometimes I feel that there is a 30-minute window and I feel like those 30 minutes can affect what the rest of your day is, based on the traffic. So, I would like to know what the fastest route is."[67] In this statement, they collapse the fastest route with the best route and reveal the affective power of Google's routing: the high of beating the traffic and the satisfaction of saving time. Google Maps allowed them to adapt to changes and fluctuations in traffic conditions and transit times, knowing immediately of any variation in conditions.

The real-time feedback that enables these traffic updates is now second nature to the Google machine; users can state an inquiry like "directions" and get an immediate response. Part of real-time's appeal is its speed–it can immediately tell you how to get to your destination quickly, because, as Kalmanowicz pitches "we all have really busy lives."[68] The omnipresent "busy lives" is echoed in the Google Maps promotional video titled *Explore Your World with Google Maps*. The video runs through a montage flurry of activity highlighting all you can do "with a little help from Google Maps." The calm voice-over prompts people to "find places you'll love," "book your table," "book your look," "get there fast" (with the voiced-over reassurance that "you're on the fastest route") and "know just where to go."[69] These videos emphasise instant information to self-sufficiently navigate these busy lives. The video presents a fast-paced, frenetic world, which Google Maps helps navigate. Real-time feedback is a means to "maintain" or keep more of one's time for the self in the name of time maintenance.

The self-sufficient explorer is in the process of "making the world your own" through collecting experiences, Google projects the impression that the world is simply there to be dominated. But these tableaus of a seamlessly convenient reality obscure the network of labour behind getting there and being there: the people who pour the coffee, clean the dishes, drive the taxis, or operate the fare booths, these perspectives are obscured in Google Maps' promotion of self-sufficiency.[70] While Fitzpatrick promises to help users "get out of a rut" by finding what's new in a neighbourhood or going on holiday, these experiences themselves rely on someone else's time and the labour of maintaining the experienced spaces. There is a social politics to the erasure of this labour, as service work is often gendered, racialised, and classed work further stratefied by legal status of immigration and citizenship.[71] Those in the service of maintaining real-time experiences are cast as what Sarah Sharma terms "temporal exceptions."[72] The inconveniences of waiting to be served or enduring a long transit route are examples of the social temporalities that facilitate the "busy lives" narrative—but Google makes marginal the people whose labour is connected to these experiences through Google's heavy focus on the self-sufficient explorer. At the same time, their labour is made hyper-visible in Google Maps through the review platform when they are seen to have slowed the Google user down.[73] In these cases, reviewers complain about slow service and sometimes call out workers by name for "taking too long," resulting in a low review. More and more, these temporalities are baked into the Google Maps interface. For example, the labour of travelling between points on the map is marked as a "15-minute" Uber ride when searching for transportation options on Google Maps. Similarly, restaurants are often awarded a glowing review because of "fast service" or being a "timely option for lunch." The temporalities of the workers who make these commutes and these services possible are made marginal to the centrality of YOU, the self-sufficient explorer who can get there fast—entitled to erase the labour while extracting others' time invested in this labour in the name of optimising your own.

Kalmanowicz's chorus of "busy lives," and "getting there quickly" presumes "making time" an individual achivement. Through a critical look at the cultural politics of time, Sarah Sharma shows how capitalist logics of saving time produce and condition a fantasy of a generalised sped-up temporality. But these logics are based on enforcing a biopolitics of time management.[74] Low-wage content managers and data taggers work long hours around the world.[75] The work of ensuring real-time persists via the maintenance of workers, those in the service of care, or those working night shifts, as a few examples. This work of maintaining real-time is often made invisible while the expectations of immediacy are more and more normal. So, while the map positions itself as taming geographies by providing the best experiences to claim and own, the people who are part of making space are effectively erased. The erasure of work happens when experience fails to be accountable to the relationships that constitute it. In short, experience and optimisation assume an imaginary universal access to leisure time and income, thus administering a biopolitics of time where the self-sufficient prototype has choices about how to move in and through the world while others don't.

Anxieties about individual "busyness" took on new urgency during the COVID-19 pandemic, when crowds and congested spaces posed threats of infection. In response, Google began to report on the busyness of sites, from stores to neighbourhoods. Before 2020, Google Maps already offered the feature of "Live Busyness" to track the number of people in a location throughout the day (by measuring occupancy through real-time location data, like how they measure traffic conditions) providing captions such as "usually busy." However, Google began to display this information more prominently on the Maps interface during the pandemic when many people sought "social distance." Google advertised this feature to "make more informed choices about the self and space, adjusting routines and habits to be more interesting, efficient, healthy, or safe," and recently expanded it to track interior areas as well as "malls, airports, and transit stations." Google Maps promises that "If it's Saturday morning and you want to explore your city without crowds bogging you down, open up Maps to instantly see busy hotspots to avoid."[76] While Google promotes this feature as a mark of safe decision-making, it fails to account for the reality that not everyone gets to curate their experience this way. Tracking this data brings into relief the privileged choice of going out into the world vs. the reality of the people working in those stores. Do they have the option of leaving if it is too busy? Like the taxi drivers, the cleaners, the coffee shop workers, and the transit drivers who maintain time, these workers are also now maintaining the capacity of the space. These narratives further detach the self-sufficient explorer from the city's infrastructures that make their experience possible. The explorer can simply benefit from the invisible work of those maintaining the space of potential experience. And while these disparities of working conditions have been called out from the start of the COVID pandemic, there is little engagement with Google Maps' role in mediating the visibility of space during the pandemic. Workers are made marginal to the Google experience of self-sufficient optimisation of one's time.

The space of Google Maps is its own extractive process. Google Maps cultivates and benefits from the aspirational labour of those who voluntarily add information to the map and moderate content that is already there. Google's Local Guides Platform creates a pool of people ready to perform the microtasks of editing business information, reporting reviews, and adding reviews. Local Guides' work is nested into the macro-tasks of the Google Maps project Ground Truth elaborated on in Chapter 2. They add in the details that cannot be obtained through satellite imagery or street view. More than a social site of sharing favourite restaurants, Local Guides are part of the human labour of fact-checking, removing millions of policy-violating reviews and fake business profiles, which they perform at the scale of hundreds of thousands and even millions of "flags" every year. Like other forms of online content moderation, the people doing this work are peripheral and made disposable in the name of optimal interaction.[77] Experience is in part reliant on the extraction of this labour to maintain the ever-replenishing resource of up-to-date local information.

Through sharing their experiences, the Local Guides Platform presents itself as the conduit of heterogeneous perspectives available on Google Maps' standardising spatial imprints; authenticating map information not just in their labour but indeed by what they represent as the so-called citizen contributor. Like social media platforms Instagram or YouTube, Google Maps extracts these forms of authenticity and realness and compensates in what Brooke Erin Duffy identifies as the"future reward systems from present day productive activities."[78] The "future rewards" include recognition from Google through prizes, promotions, and even the potential to be flown out to the Google Campus in Mountain View California.[79] Local Guides are promised the visibility of being on the map, even alerted to the number of "views" each of their review has, the full extent of their review-based labour is not always totally legible. And while Jean-Christophe Plantin argues how such forms of crowd-sourced participation on Google Maps further reify the maps' enclosure by rendering participation a form of database maintenance.[80] The Local Guide ensures the smooth running of experience, making the city more knowable and more claimable.

As much as it seems like the tech-savvy explorer can simply call upon Google Maps in an instant and find directions seamlessly, what is obscured from view is the labour activated in these moments.[81] Beyond the map, there is a global network of information workers made invisible in this process of readily available spatial consumption. In asking for directions, Google Maps users are calling on the labour of others, from the material of the hardware to what powers the software,[82] to the human "ghost work" that makes seemingly computationally automated devices run smoothly.[83] Kate Crawford and Vladan Joler's 2018 artwork, Anatomy of an AI System, details the vast global scale of materials, data, and labour activated in a seemingly simple request like asking Amazon's Alexa to turn on the lights, or in this case, asking Google for directions to the nearest coffee shop.[84] These are networks that exploit, pollute, dehumanise, malign, and yet the story that is narrated is about the person who can get from A to B without "having to stop and ask for directions." When Google Maps situates experiences mediated through the map as articulations of personal power, it renders space transactional and consumable. To command authority over experience is to not simply to take information from the map, but to activate Google Maps' global networks of data extraction.[85] The labour and the relations that make and negotiate that space are abstracted, obscured, and devalued. The leisure and efficiency of experience make such extractive demands on others justifiable.[86]

While Local Guides might seek out visibility, reviewing space on Google Maps illustrates the unequal stakes of visibility on Google Maps. In research with Aparajita Bhandari, it is not just the businesses that get reviewed but the people who work at these places.[87] This is especially true of chain restaurants where workers are made hyper-visible on the platform to people reading the reviews and to their employers. Moreover, employees often wear name tags, sharing personal information not by choice but by company policy. This information becomes weaponised on the platform where people are often called out by name and targeted for their "service" with responses from their employers. Often the reviews are acts of racist and classist targetting, that unjustly locate people on the platform without space to opt out. While the platform pushes the logic of sharing local knowledge and mobilities through various locations, the worker is made static, poised to be reprimanded.

Experience is an operation of the map—excavating the "best" a city has to offer, mining the fastest route, occupying the cleanest environment, and taking advantage of "local favourites." While explore is a spatial technology for laying claim to space, experience is a tactic of reaping all of space's rewards, at least when facilitated through the Google Maps' logics. The focus is on you—what Wendy Hui Kyong Chun describes as the "the main character in a drama called Big Data."[88] The prototype of the self-sufficient explorer remains at the centre of this imaginary. The labour of maintaining space, of making time, of facilitating busyness, is erased from view. From here, I turn to other imaginaries and realities of access that make space legible to think more broadly about what it means to make a space accessible.

**The Insufficiencies of Access**

While explore and experience are used by Google Maps as forms of accessing space—to be able to go anywhere and do anything—the normative proto-typing of these forms of exploration and experience already makes access inaccessible. Google Maps mediates care through a narrowly defined version of access that focuses on "knowing where to go" and "moving with confidence." Such suggested means of using a map can make space more knowable, predicting where to go and what to do by seeing an image of the space, getting a sense of how long to get there. Through these repeated promises and processes of explore and experience, Google Maps produces an imaginary of self-sufficiency that frames access as a method of laying a personal claim on space. But Google's branding of access is limited when considered through alternative frameworks of radical access from disability justice[89] and critical access studies.[90] Google Maps' terms of access and accessibility are based on established regulations and legal compliance with those regulations. As a result, access is templated to pre-defined terms of being in space—terms that, as I address later in this section, do not adequately meet the access needs of so many people.

Explore and experience become blanket promises often overlook or even disparage the possibilities of mutual care and connectedness. For example, in the 2019 Keyword post titled "Let Google be your holiday travel tour guide,"[91] Katie Malczyk, contributor to Google's product blog The Keyword (the de-facto press release site), relates her experience of planning a trip to Greece noting "I want to be able to navigate transportation options without having to stop and ask people for help all the time." Malczyk's framing of inconvenience makes even the low-stakes forms of interdependence (like asking for directions) a hassle, delimiting what is and what is not a disruption. In other words, Google Maps organises space around a self-sufficient user at the expense of engaging with the relationality of space. Access to space is access to Google-mediated self-sufficiency.

For Critical Disabilities scholars like Aimi Hamraie, Google Maps' terms of *access* are narrowly defined through legal frameworks of accessibility and measured by compliance to these frameworks. For example, Google Maps recently launched their "accessible Places feature,"[92] which Google defines as a way to "find out if a place is wheelchair accessible." This means that when searching for places on Google Maps, people will see a "wheelchair icon" denoting whether it has an accessible entrance. Scrolling down through the information, one can also see if there is accessible seating, restrooms, and parking. The presence of this icon is part of Google Maps' crowdsourcing effort to add information about "accessibility" to Google Maps.

Google Maps' discourses and practices of accessibility regulate and constrain access. While Google's approach to mapping access seems to be open and inter-determinate based on crowdsourcing efforts to index and label the map, Hamraie argues that Google Maps' ontology of access is limited. Indeed, Google's primary icon for labelling accessible spaces is used to mark access for mobility devices. This icon is usually added by the business or a patron of the business based on how the business meets the standards of mobility access with ramps, lifts, and accessible toilets. But as Hamraie argues, even in the specific contexts of accessibility for mobility devices, this model of accessibility leaves out other factors that make spaces accessible such as floor texture, space between aisles and shelves, or access to toilets. And while crowdsourcing might seem to present opportunity to expand these criteria of access, what Google models is what Hamraie terms a "normate template" of infrastructure that gets reproduced through the map.[93]

Google's crowdsourcing enforces what Hamraie terms a "depoliticised compliance model, which takes accessibility standards for granted as objective and neutral measures."[94] Further, Google presumes that participation means access to a large reach, and in term, vast amounts of data. However, such an emphasis on reach, as Hamraie points out, fails to contend with the quality of the information shared and the frameworks of access that people can work within, which in the case of Google Maps, is defined through the Americans with Disabilities Act (ADA). As Hamraie argues, the ADA "purports to protect disability as a category but limits the definition of disabled people to narrow, specific, and historically constituted categories."[95] Such forms of accessibility compliance foreclose the intersectional and "cross-disability" approaches to what disability justice activists call "critical access."[96] Compliance models approach disability as vsible, shared, definite, and constant. Moreover, Google's bounded calls for crowdsourcing reflect what Hil Malatino, in conceptualising trans intimacies and trans presences, calls "an affect oriented to futurity."[97] Applying Malatino's theory of "future fatigue" in trans-intimacies helps to see how the crowdsourced labour behind this mapping project is predicated on linear narratives of a future, better world.[98]

Amie Hamraie offers critical access studies as a generative lens through which to critique normative forms of access based on regulations and compliance, to instead imagine access as a dynamic relation between bodies and space.[99] Hamraie's Mapping Access initiative based at Vanderbilt's Critical Design lab is a project of adding information about access to the map based on a question of what it means to be accessible in the first place. Hamraie writes that "although this space complies with legal standards, users report that significant barriers to participation remain due to the way spaces are utilised and organised. The accessible entrance leads to stairs. Trashcans block the all-gender restrooms. The building is heavily scented (which is problematic for people with chemical sensitives but allowed under law)."[100] Adding these details to the map makes the process of mapping "interrogative" instead of descriptive. And rather than simply linear, the project asks questions about mapping process, reflective of radical disability activism. As Hamraie explains, "for radical disability activists of the 1970s, mapping inaccessible spaces was a political tool; visual evidence of discrimination established disability as a minority identity rather than a medical problem requiring a cure."[101] Mapping Access works towards what might be called a collective access.

Other alternatives to Google's accessibility mapping, such as the Radical Access Mapping Project (RAMP) in Vancouver, also take an iterative approach to mapping. Rather than a data overlay or simple icon, RAMP is an ongoing, place-based audit of Vancouver—an ever-growing list of stores, health centres, arts buildings, community places, and public toilets. Places that have been mapped include the M2M Steam Bath, Heartwood Café, Vancouver Playhouse, Vancity Theatre, Rainbow Exchange, and Little

Sisters Bookstore—revealing a focus on a range of pleasures. The audit is based on different access types related to elevation, blindness and vision limitations, wheelchair and mobility limitations, scent reduced/free spaces, and provides information on how to get there. RAMP calls for people who are providing information for this type of "mapping" to include details about signage and wayfinding, lighting, vehicle and mobility scooter parking, exterior access, doors, public transit, seating, emergency seating, water fountains, staff awareness, and public washrooms. RAMP defines their project as a way of "working together to increase access & potential through the sharing of accurate, up-to-date, easy-to-find information about the spaces we use to grow our amazing communities."[102] While acknowledging its limitations and asserting its presence as an ongoing negotiation of space, the project asserts that "it does add something, it does create dialogue and tangible change. And in a society that is consistently telling variously disabled folks that our presence doesn't have an impact, that our lives are not worth making and sharing space with, that does mean something."[103] RAMP's audits are undertaken not to enforce a compliance framework but to be a common space for sharing details about place that evades normative models of placemaking and becomes a collective means of care.

Google Maps brings into focus the "market-driven" lack of care that permeates everyday geographies and continues to place profit over people. By contrast, care networks and models of mutual aid have long demonstrated what access can be far beyond the legal compliance framework.[104] These information networks are deliberate interventions in the formalised administration of violence and harm. For example, Piepzna-Samarasinha asks "what does it mean to shift our ideas of access and care (whether it's disability, childcare, economic access, or many more) from an individual chore, and unfortunate cost of having an unfortunate body, to a collective responsibility that's maybe even deeply joyful?"[105] Of course, networks of care are nothing new. Piepzna-Samarasinha writes of care webs that are not simply at the centre of theory but have long been lived and practised realities. Google's terms of location awareness fails to contend with the complex geographies of care, preferring instead individual optimisation and self-sufficiency. Access is quantified and reduced to information management, thus depoliticising it instead of understanding oppressive barriers. As such, Google Maps is the social reproduction of the status quo. These alternative maps move away from the dispassion of self-sufficiency and towards the relations, the sensorial, the fault lines.

Mapping for connection is a feature of the Queering the MapStory Project where "People anonymously pin their experiences, memories and histories to an online global map, and each pin contains a unique micro-story of queerness. These are stories of the meaningful and mundane experiences of LGBTQI+ life. They form a living archive, collectively building new forms of community across the globe."[106] Additionally, FemTechNet, an international network of scholars, students, and artists, provide a resource to their own feminist mapping projects such as the Situated Knowledge Map which calls for people to use mapping to respond to the prompt: "Drop a pin or make a marker on a place that represents a moment of feminist knowing, unknowing, learning, unlearning, understanding, confusion."[107] Projects such as Queering the MapStory Project and Mapping Access draw on practices of participatory mapping facilitated through digital media to generously expand how access is lived and experienced.

These processes are ongoing, iterative, and take time, reflective of T.L. Cowan and Jas Rault's Heavy Processing as a method and information technology. They write, "that before you can identify your protocols, your ethics or your method, you need some heavy processing. You need listening and talking and asking for more information."[108] Projects such as Mapping Access, RAMP, and Queering the Map Story move away from the dispassion of self-sufficiency and speed; instead, they move towards relations based on interdependence and care. In this understanding, access is not the by-product of colonial exploration and extractive experiences but an ongoing negotiation that rejects at its core the self-sufficient explorer prototype for the violence it reinforces.

When organised through exploration and experience, access is a fantasy of the self-reliant prototype moving through space efficiently, cleanly, and self-reliantly. Google Maps activates explore and experience as technological features that allow one more than just location awareness but also a right to space. The hero employing these actions of explore and experience is the self-sufficient prototype—the anyone, everywhere, where all is accessible. These geographies of self-sufficiency are asymmetrical spaces contingent on white entitlement baked into the charge to "make the world your own." While they are presented as promises, *explore*, *experience*, and *access* are enunciations of imperial logics of ownership of all territory that assume a standardised interaction with the map. Moreover, as a the standard, the self-sufficent prototype operates as a means to "code and crystalise everything else as difference"[109] (to apply Mulvin's analysis) and render space outside these norms inoperable. In broadening out modes of access based on understandings of care rather than the bulldozing of self-sufficiency, access is not a mark of exclusion but a way of negotiating interconnectedness. What these critical modes of access demonstrate are the hard limits to Google Maps' view of "the universal" and how geographically insufficient Google Maps is in the face of all the radical spatial practices beyond the map.

## Notes

1 Bagil, "Google's New York Footprint."
2 The reason for this response: Canada has notoriously expensive data plans when compared to the United States. Knight, "Canada's Cell Phone Prices."
3 Noone, "From Here To."
4 Pew, "The Smartphone Difference." A 2015 Pew Research Centre study found that 90% of American smartphone owners over the age of 18 had used their

phones to get directions or information related to location. That percentage increases to 95% when considering my age bracket, 18–49. Google research from 2017 shows that one in three mobile "searches" are location related. These statistics animate the practice of using digital maps as routine, at least in the U.S.: Google for Developers, "Google Maps APIs."

5 Thiagarajan and Akasaka, "Building a Map for Everyone."
6 Google for Developers, "Google Maps APIs."
7 Find more on prototyping whiteness in Browne, *Dark Matters.*
8 Google for Developers, "Google Maps APIs." I/O is an annual conference and product showcase Google hosts at their campus in Mountain View, California. Here Google managers and executives share new software and hardware with developers. The videos are all recorded and shared on the Google for Developers YouTube channel https://www.youtube.com/@GoogleDevelopers.
9 Google Maps, "Explore and Navigate Your World."
10 Google Maps, "Why We Map."
11 Google Maps, "Why We Map."
12 Google Maps, "Why We Map."
13 I discuss Ground Truth in Chapter 2.
14 Simpson, *As We Have;* Smith, *Decolonizing Methodologies;* Tuck and Yang, "Decolonization Is Not a Metaphor."
15 Tuck and Yang, "Decolonization is not a Metaphor."
16 Simpson, *As We Have Always Done.*
17 Tuck and McKenzie, *Place in Research.*
18 Simpson, *As We Have;* Smith, *Decolonizing Methodologies,* 1–18.
19 Tuck and McKenzie, *Place in Research*, 60; Beech and Jordan, "Toppling Statues," 3–15.
20 Simpson, "Indigenous Resurgence and Co-Resistance," 19–34.
21 Dickenson, "Protesters Toss Statue."
22 Milman, "Christopher Columbus Statues Toppled."
23 BBC News, "Protesters Topple Columbus Statue."
24 Tuck and Yang, "Decolonization Is Not a Metaphor."
25 Tuck and McKenzie, *Place in Research*, 60.
26 Rault, "Window Walls and Other Tricks of Transparency," 937–960; Christen, "Does Information Really," 2870–2893. Here Rault is thinking alongside Kimberley Christens' critique of open access and information freedom as a colonial device of claiming and owning Indigenous knowledge for the benefit and control of non-Indigenous regimes.
27 Rault, "Window Walls," 948.
28 Grande, *Red Pedagogy.*
29 Tuck and McKenzie, *Place in Research,* 49; Pratt, *Imperial.* Tuck and McKenzie thinking alongside Mary Louise Pratt.
30 Simpson, "On Ethnographic," 67–80; Tuck and Yang, "Unbecoming Claims," 811–818; Smith, *Decolonizing Methodologies.*
31 Tuck & McKenzie, *Place in Research*, 66–67.
32 Tuck & McKenzie, *Place in Research*, 66–67.
33 Mulvin, *Proxies.*
34 Google for Developers, "Keynote (Google I/O '18)."
35 Google for Developers, "Keynote (Google I/O '18)."
36 Google Cloud Tech, "Going Beyond the Map."
37 Conde, "Google Touts Its Cloud Network," 6.
38 Google Cloud Tech, "Going Beyond the Map."
39 Ahmed, *On Being Included.*
40 Hoffmann, "Terms of Inclusion," 3539–3556.

41 Walia, *Border and Rule*, 15.
42 Walia, *Border and Rule*, 25.
43 Massy, "A Global Sense of Place," 149.
44 Benjamin, *One Way Street*; Marcus, "The Long Walk."
45 Tonkiss, *Space*, 23.
46 Tonkiss, *Space*, 103.
47 Cadogan, "Black and Blue," 142.
48 Cadogan, "Black and Blue," 143.
49 Fiske, "Surveilling the City," 71.
50 Maynard, *Policing Black Lives*, 89.
51 Maynard, *Policing Black Lives*, 89.
52 Sharma and Towns, "Ceasing Fire and Seizing Time," 28.
53 McKittrick, *Demonic Grounds*.
54 Singh, "Resistance in a Minor Key."
55 Sutherland, "Making a Killing."
56 McKittrick, *Demonic Grounds*, xi.
57 Weckert, "Google Maps Hack."
58 Cadogan, "Black and Blue."
59 Noble and Roberts, "Through Google-Colored Glasses," 187.
60 Google Maps, "Explore and Navigate."
61 Google Maps, "Explore and Navigate."
62 Google is strangely obsessed with tacos used as examples of consumption in several product announcements and marketing videos.
63 Google Maps, "There's More to Explore."
64 Massey, *For Space*, 29.
65 Sharma, *In the Meantime*.
66 This interaction took place in Toronto, Canada in the autumn of 2017 as part of *From Here To*.
67 This exchange is paraphrased from one interaction part of "From Here To" from the Autumn of 2017.
68 Kalmanowicz, "Make the World."
69 Google Maps, "There's More to Explore."
70 This is a central focus of Sharma, *In the Meantime*.
71 van Doorn, "Platform Labor," 898–914; Block and Hennessy, *"Sharing Economy;"* Walia, *Border and Rule*.
72 Sharma, *In the Meantime*, 57.
73 Bhandari and Noone, "Reviewing 'Local' Experiences," 198–207.
74 Sharma, *In the Meantime*.
75 Gray and Suri, *Ghost Work*.
76 Phillips, "Maps Is Getting More Immersive and Sustainable."
77 Gillespie, *Custodians of the Internet*; Roberts, *Behind the Screen*.
78 Duffy, *(Not) Getting Paid*, 70.
79 Bhandari and Noone, "Support Local."
80 Plantin, "Google Maps as Cartographic Infrastructure," 489–506.
81 West, "Censored, Suspended, Shadowbanned"; Graham and Dittus, *Geographies of Digital Exclusion*; Roberts, *Behind the Screen*; Gray and Suri, *Ghost Work*.
82 Hu, *Prehistory of the Cloud*; Hogan, "Data Flows"; Hogan, "Big Data Ecologies," 631; Crawford, *Atlas of AI*.
83 Gray and Suri, *Ghost Work*; Roberts, *Behind the Screen*; Gillespie, *Custodians of the Internet*, 207–208
84 Crawford and Joler, "Anatomy of an AI System."
85 Extraction reflects Kim Reynolds's concept of "extraction as white supremacy" which presumes a uni-directional relationship of taking and owning. Reynold's

concept of extraction is embedded in Google's investment in programmes like Next Billion Users and The Google Station—its internet infrastructure building project that targets cities in India, Mexico, Indonesia, the Philippines, Thailand, and Nigeria—to "mine the market potential" of the Global South. These colonial projects not only target the idea of Google users but also of Google workers available to add and moderate spatial information as well as expand Google Maps' coverage of experience. While I discuss these projects in greater detail in other chapters in this book—Chapter 5 looks at Google Maps' coding of absence and error, and their projects of Plus Codes and StreetView mapping of favelas in Brazil—I reference these projects here to mark how they undergird Google's mythology of the self-suffcient prototype and their quest experiencing the world.

86  Sharma, *In the Meantime.*
87  Bhandari and Noone, "Support Local."
88  Chun, *Updating,* 53.
89  Piepzna-Samarasinha, *Care Work.*
90  Hamraie, "Mapping Access," 455–482; Hamraie, *Building Access.*
91  Malczyk, "Let Google."
92  Google Help, "Find Wheelchair-Accessible Places," 7.
93  Hamraie, *Building Access.*
94  Hamraie, "Mapping Access," 464.
95  Hamraie, "Mapping Access," 466–467.
96  Hamraie, "Mapping Access," 455–482.
97  Malatino, "Future Fatigue," 636.
98  Malatino, "Future Fatigue," and Malatino, *Trans Care.*
99  Hamraie, *Building Access;* Hamraie, "Mapping Access."
100 Hamraie, "Mapping Access," 466.
101 Hamraie, "Mapping Access," 455.
102 Radical Access Mapping Project, https://radicalaccessiblecommunities.word-press.com.
103 Radical Access Mapping Project, https://radicalaccessiblecommunities.word-press.com.
104 Spade, "Solidarity Not Charity," and Spade, *Mutual Aid.*
105 Piepzna-Samarasinha, *Care Work,* 33.
106 Kirby et al., "Queering the Map," 1043–1060.
107 Cowan, Surkan, and Wexler, "Situated Knowledges Map."
108 Cowan and Rault, "Heavy Processing, Part 1."
109 Mulvin, *Proxies,* 89.

# 4    Orientations of Legibility

**Templates | Surfaces**

I start in Amsterdam, outside the Rijksmuseum, looking for directions to the Albert Cuyp Market. The direction-giver initially spoke their direction; but, when I asked them to draw, they pulled out their phone and offered, "here I'll show you," presenting me with Google's visual representation of the city. They did not use Google Maps' search function to find a route. Instead, the map's blue dot visually established where we were standing, where I was to go, and the space between the two. Rather than draw, they narrated the pathway I was to take, using the map image to coordinate meaning. First, they located us with the remark, "we are the blue dot" and, from there, indicating which way I was to go, dragging their finger along the screen. They said they wanted to "show me" the directions using Google Maps rather than draw the directions because Maps' birds-eye view was, to their mind, easier to understand and remember, without the confusion of their interpretation and memory of space. Google Maps made the city in-between *here* and *there* navigable and legible rather than obscured by their impression.

The city is a central figure in Google Maps' imaginary of location awareness. Google Maps' About Page is a rich visual display of potential encounters, peppered with images such as a person walking through a narrow table-lined street in Rome, to friends queuing for coffee at a bustling café in New York City, to gridlocked cars cemented to the lattice streets of Chicago.[1] These snapshots of city experiences are captioned with the promises to *make plans happen, know how busy a place is,* and *receive real-time traffic updates*[2] —overlaid with a screenshot of the Google Maps interface, illustrating how to put each command in action. The map's placid precision of pinned locations, set on clearly outlined streets, offsets the city's complex circuitry, full of congestion and commotion. In other words, Google makes the messy city a legible place.

In many ways, Google's claim of organising spatial information is most compelling in places like cities, popularly portrayed as overloaded with information: bursting with people and stimuli. Symbols of urban density such as Times Square in New York City or Shibuya Crossing in Tokyo, Japan, two

DOI: 10.4324/9781003251569-4

of the world's busiest intersections, are iconic for their bustling crowds and pulsing flows. A search on Getty Images yields hundreds of thousands of licensed photos of Times Square and Shibuya Crossing, with each image containing a surge of activity caught in the split second of a photographic frame. These renderings of life—a sliver of the city's intensity—evoke a heavily concentrated informational space. Cities are host to a myriad of sensorial cues and rhythms from the clusters of people moving through the streets, the hums of their passing conversations, the chirps of pedestrian crossing signs, the sparkle of advertisements, and the diffuse lights from the windows of the towering buildings. Layered on this information landscape are the declarative street signs, the directional commands painted on roads, or the new parking rules affixed to lamp posts with zip ties and Sellotape. Add to that shortcuts, speed traps, tram tracks, road works, detours, and low emissions zones. Many of these cues help in reading space; but, together, these can feel like a wad of sensorial commands and directions.

On top of the bricolage of wayfinding information assembled in the built environment, the city is filled with networks of digital information that we can access via mapping platforms to trigger location awareness in a moment's notice. Rob Kitchin and Martin Dodge theorise this enmeshing of digital information with the built environment as "code space," to reflect the often-hidden pervasiveness of datafied systems.[3] We might activate these forms when tagging locations on Instagram, calling up a ride-share car, or reading a restaurant review online. The tacit informational and infrastructural networks holding place together are what Mattern refers to as the "ambient intelligence" of computational and non-computational data processes working behind the scenes at all moments collecting, processing, transmitting, and storing.[4] Mattern reminds us, these digital networks are visible through the signal towers on top of tall buildings but also less visible material networks like fibre optic cables and hidden away data centres.[5]

Google's portrayal of city life evokes the calming and containing of a metropolis's information, reflective of its promise to make the world "understandable and accessible" while at the same time engaging the many experiences that make places desirable and alive.[6] The promise of locative media like Google Maps is to organise the immensity of city intelligence into a recognisable aesthetics—a collage of information, layered with the impressionism and abstraction. And while maps have long been tools of city governance, used to manage the complexity of the metropolitan environment, Google Maps diffuses the ideal of a legible system through the variability of a dynamic map. Google's dynamic map is one of the layers of information built and orchestrated through projects like Ground Truth (discussed in Chapter 2)—projects premised on collecting the totality of information. Google Maps' representational quality is more than a mimetic transfer of space, but also a space animated by the environment's affects defined through Google Maps' features such as *area busyness*, *live traffic updates*, *the fastest route*, or even *air quality*. The question is how does one form an impression of place that is both a single moment and an enduring sense like the impressionist painters of the 19th century? What from the flux is fixed, even if held within the promise of a dynamic map?

For Didem Özkul, digital maps impose an "algorithmic fix" to place-making, mobilities, and location by attaching behavioural inferences to locational data.[7] These become "predictive, pre-emptive, and discriminatory in how they 'fix' who we are and whom we may become with the aim of creating a predictable future."[8] This indexing of space sets how space and people are made legible to each other. Here Özkul moves between the dual metaphors of fix in terms of fastening and mending. The algorithmic fix fastens one to a location or position while also claiming to be a go-to solution for navigating an information-saturated landscape. Here Özkul builds on David Harvey's application of the spatial fix[9]—how space is reorganised to make room for capital through imperial process—to be about how locative media, by way of algorithmic governance, overdetermines one's location in the world, be that socially or geographically.[10]

Locational fix operates in tension with spatial flux—the indeterminate way by which space unfolds beyond representations, however static or dynamic the representations claim to be. This unfolding of space reflects what Nanna Verhoeff terms the "performative cartography" of wayfinding and place-making with screen-based interfaces that complicate the "visual regimes" of navigation.[11] Verhoeff writes, "the visual regime of navigation, that which is depicted, such as maps and panoramic views, emerges simultaneously with someone's interaction with a screen-based interface. This simultaneity of making and image makes movement itself a performative, creative act. Movement not only transports the physical body, but it also affects the virtual realm of spatial representation."[12] While Google Maps may fix space to its vernacular of representation, one's reading and engagement with Google Maps informs their own processes of making the city legible that works with and sometimes against Google's imposed logic of space.

The chapter traces some of the leaks and cracks in the Google Maps edifice of legibility. In considering how Google's mapping project attempts to stabilise and coalesce spatial meaning-making, this chapter frames Google's project of legibility as a project of making space *appear* governable. Even perforated attempts to attend to space as dynamic and immersive flatten space to a simple surface. These competing fixes and fluxes of Google's terms of legibility reveal how legibility models a reading of space beyond navigation. In drawing attention to the punctures in Google legibility, the goal of this chapter is not to fill the holes but instead to stand in the cracks of Google's fractured fantasy *that space can be held*, and to outline other contours of *being in space*.

## A Template for the Legible Map

To be legible can mean to be readable and comprehensible—a quality of both form and content. Curiously, legibility is itself a variable condition of communication—implicitly inter-determinate, demanding both an author and a reader. Disciplines like design and education have endeavoured to measure what makes something legible in order to develop protocols that coordinate sensemaking. Such bids to define legibility often do so through standards of measurement, imposing quantitative values on qualitative phenomena. In handwriting, the Handwriting Legibility Scale (HLS) measures legibility on a scale from 1 to 5 to assess the "overall impression" of handwriting measured via the criteria of speed, the number of words produced in six minutes, to spelling, letter formation, to the layout of the page.[13] Similarly, the Flesch Reading Ease Readability Formula is a similar system used to determine how easy a text is to read, measured as *Readability Ease*, based on a scoring system of 0 to 100.[14] The Flesch Formula assesses factors such as the average number of syllables per word and the average sentence length to assess reading facility. That which meets the requirements of word length, font type, and comprehension are presumed to be legible. Both the HLS and the Flesch orient reading practices according to a fixed set of rules and standards to coordinate the complexity of legibility's multivalence.

Translated to cartography or mapmaking, legibility is also a balance of form and content in relation to the practice of translating a three-dimensional space to two-dimensional surface. As Arthur Robinson argues in his canonical *Elements of Cartography*, cartography is not meant to produce a mimetic copy of the work but a "chart to be worked on."[15] Questions of scale and projection are met with other questions such as: is there enough contrast between light background and dark text? Is there enough detail to inform but not too much detail to confound? Are the relations between label and object apparent? The question of legibility is not one of accuracy but of diagrammatic legibility.

Jacques Bertin's *Semiology of Graphics: Diagrams, Networks, Maps* is a theoretical exploration of meaning-making through the visualisation of information.[16] For Bertin, "It is obvious that: the most efficient constructions are those in which any question, whatever its type and level, can be answered in a single instant of perception, that is, IN A SINGLE IMAGE."[17] The treatise, the result of his work as a cartographer and geographer, establishes and verifies not just the forms of representation itself but the epistemological commitments undergirding static representations of space. For Bertin, applying symbolic and diagrammatic logic to the visualisation of information needed to follow a set of principles and frameworks in order for it to effectively convey meaning and to make spatial projection legible.[18] As Matthew Edney notes, Bertin routinely updated his trestise on graphics, with small changes to every publication, demonstrating the dynamic and relational tendencies of legibility. But ultimately, the thrust of this text was to arrive at or unearth a common and formalised legibility that tempered and evened out variables such as time, composition, and context that inform practices of reading space.

The tension between the map's forms and functions, the fixity of form and the flux of reading space, evokes the question offered by Eduardo Camacho-Hübner, Valerie November, and Bruno Latour: "Do maps and mapping precede the territory they 'represent,' or can they be understood as producing it?"[19] In other words, even if the forms are "worked on," then these images are nevertheless, at the moment of presentation, forms through which to read and interpret the world, or at least be oriented towards a particular reading. Designing maps to be readable and usable can be a process of bending the geographic truth to make sure that labels are readable, in turn establishing how one reads the space it represents.

The London Underground Map, or the Tube Map, is an example of a map designed for legibility rather than geographic representation. The Tube Map that is used today is based on a design by Harry Beck in 1931,[20] expanded to include other parts of the London Transport Network that developed since Beck's interpretation such as the Overground Network, the Docklands Light Railway, and the Elizabeth Line. Beck's map of the London Tube was a response to the previous maps of the underground system that focused on representing the lines and stations of the network according to the terms of geography and distance. Geographic representations of The Tube looked crowded in the central area of London where the lines converged, resulting in a style of representation that made the Tube Map difficult to read. In contrast, Beck's map was a diagrammatic interpretation of the system, premised on producing a map that made all of the lines and stations visible, including the labels of each station. The Tube Map helped make the knotted streets of London and its arteries of transit seem navigable.[21]

To maintain clarity of the map meant prioritising the legibility of diagrammatic organisation over the mimetic representation of the underground system. In other words, the positioning and scaling is a bit off. One example is that Embankment and Charing Cross, consecutive stations on the Northern Line, appear as distinct stations, spaced at the same distance allocated between many of the stops on the Northern Line. On the ground, these stops can be accessed within the same block, less than 100 meters away from each other.[22] At the other end of the spectrum, the distance between London Bridge Station and Borough Station on the Northern Line appears like one of the longest stretches between stations, which in relation to other distances, would signal a significant walk. However, in practice, it is only about half a mile between stations and could be bridged with a ten-minute walk or a less than three-minute Tube ride. So, while maps help govern the complexity of systems, and make movement through the city via the underground imaginable and legible, these maps impose a reading of space that may be abstracted from the actual conditions. Decisions of design are themselves spatial transformations that as Johanna Drucker argues, become generative of knowledge.[23] These, in turn, orient their own forms of location awareness. Visual systems like the London Tube Map, translate processes of legibility to be "reality" rather than "representation" but even more so, the map and the city are themselves reread and reperformed with every interpretation.

Janet Vertesi's research about visualising London illustrates how Beck's iconic map does not simply make the Underground newtork legible via 2D representation, but comes to inform an imagination of the city itself.[24] Vertesi asked people she encountered in London for directions, and then asked that they draw these directions. Vertesi found that in asking people to draw representations of London, they incorporated drawings of the Underground stations into their depictions, often interpreted as landmarks, and arranged these stations according to the logic of the London Underground map rather than the topology of the streets. Vertesi remarks how "the Underground bled into the above-ground in their representations of London."[25] Indeed, the London Tube Map often informed how people visualised their movement through the city, even if they were not taking the tube. Vertesi's findings illustrate how popular representations of territory cajole and condition a reading of territory to "tame and enframe" complicated realities, such as the London streets system that in turn produce their own readings of space.[26] As Vertesi argues, the iconic nature of the London Underground image means that the mediating technology is more than simply the map, but "the image itself."[27] So while the Underground map is not designed to be mimetic, it nevertheless is read as a direct representation.

Google Maps reproduces a practice of making the city and its networks legible through its now familiar visual language of mapping. It uses the same colour scheme and grid transposed on all cities, presenting space according to a prescribed colour palette: the streets are white, hospitals are pink, parks are green, "areas of interest," or what Google terms commercial areas are yellow, water is blue, and residential areas are grey. There are the familiar "pins" that indicate precise locations: orange "pins" for commercial restaurants and grey "pins" for a nearby church as well as a health service centre and blue squares denoting transit stops. All text is in Arial font, labelling streets, sites, and neighbourhoods. While the specificity of these forms may slightly change over time as Google updates, Google's ordering of space is not only familiar but the dominant visual vernacular of digital mapping. And even what Camacho-Hübner, November, and Latour described as the "shock" of zooming in on the map of encountering an advertisement for a fast-food chain or a big box store is now a familiar part of looking at Google Maps' interface.[28]

Part of what makes Google's organisation of space familiar is that the visual protocols of its representations are transferable from one city to another. Oren Naim, Director of Product at Google Maps, even goes so far as to promise the same familiarity of coverage from "Tokyo to Tonga."[29] Looking at these two places on Google Maps, we see how the yellow lines mark out major arteries, white lines for other streets, blue pins marking places to buy groceries, flat greens, flat greys; in short, both places look like somewhere I could visit and know. All cities, while they may have vastly different layouts, follow the same visual cues, in themselves made all the more familiar. Zooming in and out of different cities reveals a standardising visual language of Google's mapping, applied to everywhere it maps. These are more than simply conventions carried across different maps; rather it is one very large map that is at once interface, global dashboard, and spatial template unified through Google Maps.

Doreen Massey argues that maps in general treat space as a surface, informing how one relates to space as that which is "continuous and given."[30] But as Massey notes, the crucial characteristic of space are that space cannot be "reducible to a surface."[31] Maps are produced through making choices about representation. In the name of legibility, this means that things can be simplified, or flattened, while other things are just plain left out. Even topographical depths, while often accounted for through a concentration of lines and markings, abstract the affects of elevation from changes in breathing, to the fatigue of a steep incline, to the vistas only available at these heights. Moreover, spatial reasoning is not exsluively visual, but can also be tactile, sonic, and olfactory.[32] Therefore, map-as-surface is one way to think of the map's missing depths of experience.

Architects Clara Wong, Jonathan Solomon, and Adam Frampton created a map that added depths to the surface reproduction produced through Google.[33] Wong, Solomon, and Frampton's representation of space attends to the elevation levels of the city as well as other elements of a city's verticality like its bridges, walkways, high rises that one travels through to get around the city. What the authors found was other forms of mapping Hong Kong literally made it appear flat when in actual fact Hong Kong is a city built with widely varying elevation levels. But the lack of elevation is not an accident. Like Google Maps, what Wong, Solomon, and Frampton produce are axonometric diagrams of space that are designed to show multiple sensemaking points of one area of Hong Kong. This map-as-exploded-diagram contains a multiplicity of spatial annotations that produce a network of pedestrian routes through Hong Kong all travesable above ground. The's purpose of this map's legiblity is to show overlap and complexity with the city's terrain.

City streets are an important part of how Google Maps makes space readable— part of looking for directions, knowing traffic conditions, armchair travelling with Street View, or to simply checking a Street View image to have a better idea of the street's layout before going to a new place. Even as Google Maps' interface changes over time, streets remain a defining feature from which they continue to add more and more detail, from the width of streets to the directions of crosswalks, to the location of traffic lights. Google identifies and classifies this information for the millions of its street-view images. According to Russell Dicker, Senior Director of Product at Google Maps, "the sidewalks, curbs, and traffic lights look similar in Atlanta and Ho Chi Minh City—despite being over 9,000 miles away."[34] Maps' model of readability is translatable across cites, where, as Dicker notes "the same model works in Madrid as it does in Dallas, something that may be hard to believe at first glance." For Google, streets are a commanding feature of the city and a crucial element of the city's network that can be translatable to Google's vernacular of seeing space and their technical processes of templating representation.

Templates are a key part of industrial production that assume a universality and scalability of formal principles. They are instruments of standardization used to guide and shape, often applied in craftwork, woodworking, ship-building, moulding, pottery, or metal work, as a few examples. Templates, like standards, not only make norms but also make norms easily reproducible and transmittable and as such, regulate design as well as the world they are representing. Graphical interfaces are other types of templates that system-atise graphical forms or organising information. Even before the scaling of the industrial revolution, graphical templates were a key part of the print revolution, used to coordinate visual ways of processing knowledge. As Drucker writes "Beginning in the fifteenth century in Europe (and considerably earlier in Asia), printed images in scientific and technical publications created stable, repeatable, shared references, initially through woodcuts. The advan-tages of printed images were twofold. They could be circulated to create common understanding and visual reference, and they had particular graphical properties imposed by the technologies of the media in which the lines were produced and reproduced from wood and then metal."[35] Graphical templates become a language through which the world is read and understood, formalising representations to that which can be read. Templates, like standards, not only make norms but also make norms easily reproducible and transmittable and as such, regulate design as well as the world they are representing.

Google Maps' template prioritises streets and roadways, a feature of spatial legibility impressed upon all of its maps. Street names index and orient Google Maps' turn-by-turn directions.[36] But even if streets follow the same design, that is not to say they are read the same way on the ground. For example, in reperforming *this way brouwn* I noticed how the Google Maps street labelling did not match the local vernacular of space.[36] In London when I was by Liverpool Street Station asking for directions to Brick Lane, the people I approached used Google Maps to guide me. Based on the Google Maps instructions, they directed me to turn onto "the A11" and walk along "the A11" until I reached my destination. When following the directions on the ground, I could not find any signs for A11 even though it was the street name that appeared on Google Maps. As I learned after, the A11 is a major trunk road that travels through London, following different naming conven-tions in different boroughs. By Liverpool Street Station, the A11 is the Whitechapel High Street and signed accordingly. Further east, the A11 becomes Mile End Road and then Bow Road. Being directed by Google's street name was not an efficient means to navigate in this moment. While the directions appeared easy enough on the Google Map interface, the translation to the ground revealed a different organisation of space. Google's logic of the one-name street meant that Google directions did not coordinate with the on-the-ground navigational aids like street signs and local maps. In short, street names are not always central to the navigation of space, and street naming conventions are constructions of locally held knowledges often omitted from Google Maps' interface.

In considering the signification of legibility and illegibility, Pepita Hesselberth et al. consider the limitations of the hyperlegible, writing that, "the hyperlegible, in turn, could be used to denote something of which certain elements are taken to be particularly easy to read in a predetermined manner according to fixed frames of legibility, leading to it only ever being read in this way (or, in the case of the a-legible, to it not being read at all)."[37] The decisions of what makes something legible—or, alternatively illegible—are culturally, historically, and contextually constructed and specific. In determining map legibility via the centrality of the streets, Google presumes a stability and universality to streets not simply as arteries of movement, but as themselves stable reference points. Reading Google Maps as a template helps to understand how its map generates, reproduces, and naturalises spaital norms as location awareness.

Michael Gilmore[38] argues that the centrality of American settler systems of roadways informs a fantasy of an organised city. In writing about Willian Penn, one of architects of American colonial settlements, Gilmore writes that, "Penn's most influential contribution to American legibility lies elsewhere: in the recording and patterning of space."[39] For Penn, Philadelphia was a city where "every piece of land is to be allocated, measured, and numbered."[40] The hard lines and angles of gridded streets can make the project of settler colonialism appear logical instead of violent.[41] Penn's modelling of cities comes through in other American cities such as Manhattan's famous grid pattern (at least starting from Houston), with interlocking lettered and numbered streets and avenues. Latticed streets are evident in other American cities like Chicago, Washington DC, and beyond. This type of legibility extends to cities like Salt Lake City, Utah, where streets are named and identified according to the "distance and direction from the Mormon Temple (Fourth East Street, Second West Street, etc.)."[42]

The question is not simply that these maps-as-templates become a model of reading as a categorical truth but also these models can be and are often nefariously constructed often for the purposes of domination and ownership.[43] For instance, the street names in settler cities like Toronto become more and more estranged from their colonial framework. Linda Tuhwai Smith aruges, street names like Queen Street and King Street, popular street names throughout settler Canada, make banal settler occupation, as simply names on streets that appear in the map, repeated in forms of directions. But these names have their own histories of erasure.[44]

The naturalised settler legibility embedded in Google Maps is problematised by interventions like the *Ogimaa Mikana Project*, by Anishinaabe artists Susan Blight and Hayden King to "restore Anishinaabemowin Placeness to the streets, avenues, roads, paths, and trails of Gichi Kiiwenging (Toronto)."[45] Blight and King re-inscribed Anishinaabeg names over settler names on Toronto's street-level signposts and added further street signage. For instance, they renamed part of Queen Street Ogimaa Mikana (leader's trail) with a sign marking the change "in tribute to all the strong women leaders of the Idle

No More Movement,"[46] further noting that "the project hopes to expand throughout downtown and beyond." In Parkdale, they added The Dish with One Spoon Wampum belt as a street sign at the corner of Queen Street and Noble, explaining the agreement first made with the Anishinaabeg and Haudenosaunee that English and French settlers were brought into. The project returns Spadina Road to "Ishapadinaa," and Davenport Road to "Gete-Onigaming." The Ogimaa Mikana signs are ways of asking, "how do you recognise it [this place]?" and compelling passers-by to think about how and what we recognise in our cities, and how this recognition is based in settler logic.

Sarah Sharma asks: "whose routes are valorised?" in the locative media project of making space legible and worlds knowable.[47] While *Ogimaa Mikana Project* does not specifically address Google Maps, it destabilises the settler logic baked into what gets mapped and takes into account what Edward Said terms the information "outside" the hegemonic frame.[48] It thereby reveals the logic of Google Maps' ostensibly universal template by drawing attention to what is prioritised (settler street names) and what is left out (Indigenous names and the relations to place). The project lays bare that which is otherwise taken for granted in the city itself. Standards of legibility are more than simply ordering presciptives, they are practices of obscurement reproduced in banal and brutal ways.

### A Surface for the Legible City

The fantasy of a legible city precedes the map. Famously, Baron Georges-Eugène Haussmann's reconstruction of Paris in the mid-19th century was undertaken in the name of transforming the tangled, densely populated, and deliberately disorienting medieval streets of Paris into wide, straight boulevards and architectural harmony, lined with gas lamps to make sure the city was readable even at night. Haussmann's goal was to produce a governable city through an architectural aesthetic of cohesive grandeur where streets and buildings would follow a set of standards. But this legibility was deployed at the expense of those who were forcibly removed from their neighbourhoods, mostly those already forced into impoverished conditions.[49]

Similar principles of standardisation and manageability undergirded Lucio Costa, Oscar Niemeyer, Joaquim Cardoza, and Roberto Burle Marx's project of redesigning the city of Brasília, Brazil in the 20th century.[50] Brasília, a planned city, is famous for its geometric street design, emblematised by wide, straight streets, cross-axial design, and superblocks. Both the rebuilding of Paris and Brasília projects are reflective of the colonial European prerogative to tame space through the presumed logic of grids, interconnected roadways, and square, organised structures.[51] Both Haussmann and Costa's city design presume the city to be a singular project that can be made comprehensive and total through design. These visual standards do not simply codify what the built environment looks like but attempt to calibrate the everyday practices of city life to their terms of legibility—uniform, direct, and totalising.

The modernist practice of managing the city and its circuits of daily life became a question of how the individual *reads* a city in the course of their everyday movements through it. In the 1950s and 1960s, urban studies scholar Kevin Lynch framed legibility as the "visual quality" of the city.[52] Lynch frames legibility as "the ease with which its parts can be recognised and can be organised into a coherent pattern. Just as the printed page, if it is legible, can be visually grasped as a related pattern of recognisable symbols, so a legible city would be one whose districts or landmarks or pathways are easily identifiable and are easily grouped into an overall pattern."[53] Lynch contended that a person's ability to picture their city and visualise their movement through the city was a sign of successful planning. In his landmark book, *The Image of the City*, Lynch outlines how wayfinding is linked to what he calls the "imageability" of city landscapes. For Lynch, imageability is how a city can be "imaged" or visually represented by the people who inhabit it.

Lynch based this theory on the culmination of a five-year study in which he and his research assistants appealed to passers-by for directions. In the streets of three American cities—Boston, Newark, and Los Angeles— Lynch and his assistants would approach strangers with the questions:

- "How do I get to ____?"
- "How will I know when I am there?"
- "How long will it take me to walk there?"

They asked participants to draw maps of their directions. The responses elicited from these prompts were supplemented by a series of on-the-street interviews about the drawings and one's internalised location awarenss.

Lynch also investigated the drawing of spatial memory in a lab setting with extended interviews, in which participants spoke about the map they drew, the choices they made in drawing the map, their interpretation of space, and how they imagined it. For example, during the Boston interviews, participants were asked, "what symbolises the word 'Boston' for you?" and then were invited to draw a map of the city. In another phase of the research process, a small number of participants were asked to draw detailed directions and to describe the sequence of events that one should follow to reach the destination successfully. Lynch also asked participants to describe emotions they may have felt in relation to specific areas of the map, such as confusion or frustration, and asked them to identify areas they had trouble representing. By taking the study from the streets to "the lab"—from the highly variable to the presumed controlled space—Lynch was after a sociological "truth" about spatial experiences in the name of fully and objectively "knowing" the experiences of city life.

While Lynch's study established an innovative means to think about city planning as a process that should incorporate and reflect the experiences of the people who inhabit that city, it nonetheless focuses on producing models to design what he calls an *optimal city*. Lynch sought a universalising concept of legibility that harmonised the variations in experiences, producing a taxonomy of formal features of paths, edges, landmarks, nodes, and districts. For Lynch, relations among the city's formal features produced legibility; and, therefore, it mattered how landmarks and paths connect and relate. For instance, does a road lead to the landmark? Or does the landmark get in the way of the road? Lynch's construction of people's perception of a city is part of how people use the city's forms to move through it, and these forms must be cohesive.

In an attempt to create the best readable city, Lynch standardises relations to space. But this is to assume that relations are direct and discrete transactions and not produced through variations, differences of interpretations, or fleeting moments. Variations of relationships to forms and between forms are evident throughout Lynch's mapping project where people who live in the same city. One such form are *paths*, which Lynch defines as "the channels along which the observer customarily, occasionally, or potentially moves" or the "predominant elements in their image."[54] Other forms include the edges, which work in "holding together generalised areas" as well as districts that one is either inside of or outside of. There are also nodes, "the strategic spots in a city into which an observer can enter," and finally landmarks or the "frequently used clues of identity."[55] For Lynch it was important to consider how these forms structure the image of place.

Lynch's project of asking for directions may seem somewhat familiar to those who recall Stanley Brouwn's *this way brouwn* project from Chapter 1. Around the same time as Lynch was asking for directions in the United States, Brouwn was walking the streets of Amsterdam, carrying with him blank pieces of paper and a pen. Along his way, he approached other pedestrians and asked them for directions like how to get to Dam Square or City Hall and then asked for this to be drawn. Brouwn would perform this action over and over, amassing a collection of hand-drawn directions; each stamped with the project's title *this way brouwn*.

The drawings of directions Brouwn elicited are not a complete nor a total map, nor are they a singular representation of the world. Like Lynch's project, these drawings translate the city's forms and movements to a series of lines, nodes, Xs to mark the spots, arrows, boxes, and scratches. But diverging from Lynch, in some way, their decontextualised lines and hard-to-read scribbles suggest a futility to the idea of a total legible space. In fact, so many of these maps are difficult to decipher. Where is that line headed? What is that street name?[56] For Brouwn, these directions are neither good nor bad. Rather than an exhaustive database of possible pathways, these drawings are about how difficult it is to index the relationality of space. Brouwn suggests the artwork "makes people discover the streets they use"[57] and to get people out of their own spatial scripts and consider new routes.

But these connections are transient since asking people to translate space may become illegible beyond that encounter. Curator Claire Lehmann describes the drawings as capturing "the human-centred experience of our surround" but also about "displacement."[58] In short, unlike Lynch's approach, Brouwn's hand-drawn directions don't amount to a complete map but instead make a type of spatial score that references a fleeting encounter of asking for directions, momentarily held between two or three people. In this context, Brouwn's work appears to be pushing back against such measurable types of city legibility that orient the city towards singular visual language of space.

Though separated from its intended function (at least from the perspective of the inscriber), the drawings are evocative in their own right. The undesignated lines that run into other lines and then have no end. The lines begin and end abruptly, jutting into other lines, dissecting, intersecting, and scribbling over. Often there are arrows and sometimes there are drawings of key points, particularly striking buildings or a bridge or a line of train tracks. They resemble the conceptual aesthetic of a Sol LeWitt line drawing or an Agnes Martin grid. Each one of the multiples represents a visual truncation of experience that undermines the inherent utility of the map, as tool. Indeed, there is a sense of freedom in the open narratives made possible through this abstraction of the city.

In reperforming Brouwn's work of asking for directions, I became attuned to many of the city's lines and markings that animated legibility. These ranged from notable landmarks and tourist attractions to recognisable commercial sites, interlacing the *Starbucks on the corner*, the *McDonald's arches*, with *the Van Gough Museum*, and *the CN Tower*. People often annotated these landmarks with qualitative descriptors of their location awareness. In other words, these were not static nodes on a piece of paper, but these were intersections introduced in contexts of other perceptual or emotional cues. These could be affective means to describe features and elements of a city like a *hectic intersection* or *my favourite restaurant* or *this way has more trees*; or it could be a means to flag a steep incline or *many stairs*; it can also be temporal awareness of location—such that a destination is perceived to be far away or close by. Sites are marked by personalised textures, composed of material and immaterial forms of attentions that serve to orient, assure, and assert a punctuation or an affectation in the landscape. The choice of marking declares a personalisation of space—the particularised sense of the city that disrupts the stable representations of space—the scruffy triangles, the library with the lions, the long walks, and the weaving fish hands that snake through an imagined sense of place. These are key markers of how to get from A to B based on discrete materialisations and sensations of legibility (Figure 4.1).

The pathways themselves were not always streets or traffic patterns. Instead, pathways were also waterways like rivers, canals, and lakes, or they could be shortcuts across public squares. I was told to *follow the river, walk along the canal, or head towards the lake*. The currents of London's Thames River and the Amsterdam Canal system, and the boundary waters of Lake Ontario and the Hudson are themselves part of the broader imaginary of these cities, and they unsurprisingly become baked into the navigation of

(a)                                    (b)

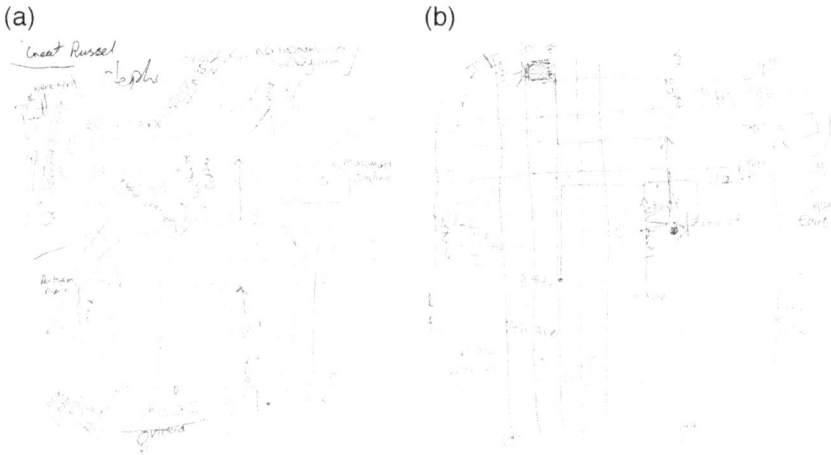

*Figure 4.1* Composite of directions from London (left) and New York (right).
Collaged by author.

space. But what sometimes augmented someone's location awareness were the discursive and metaphorical pathways that animated directional commands that gave shape to a city's tacit flows. A street might be referred to as the "busy street" or "the one with the bend" as a way to differentiate it from another street. I was often told to "follow the flow" as a means to mark out space and the direction I was to head. Flow often alluded to a type of spontaneous or unplanned action: the unmanageable and unpredictable types of flow that were assumed to exist when giving directions. Flow was based on following the movement of people, following the path of objects such as trams, or going along with an unfixed path. Pathways were more than just streets but also the tacit forms of a city, or the types of pathways one did not walk on but nevertheless could be guided by (Figures 4.2–4.5).

These elements are constitutive of the city's streetscape, or the dynamic elements of space rather than the static constructions of the streets themselves. Thinking about the streetscape opens up the creative possibilities of legibility, beyond standards of reading but as practices of *being in the moment*. This practice of *the moment* takes hold in Georges Perec's *Attempt at Exhausting a Place in Paris*, Perec's experimental novella based on observations from his seat at a café in Paris.[59] In his attempt at exhausting, he is transcribing what he sees, what he bears witness to. Every day, over and over, recording numbers in an attempt to make that specific space legible, knowable, representable. In this case, it is clear that this is a singular perspective. Ben Highmore describes Perec's work accordingly: it "isn't just a valiant effort to attend to the insignificance—it also registers the necessary contingencies involved in this."[60] Part of the joy of reading Perec's account *is* the limitations—the absurdity of the inventory. Perec, laying out the precincts

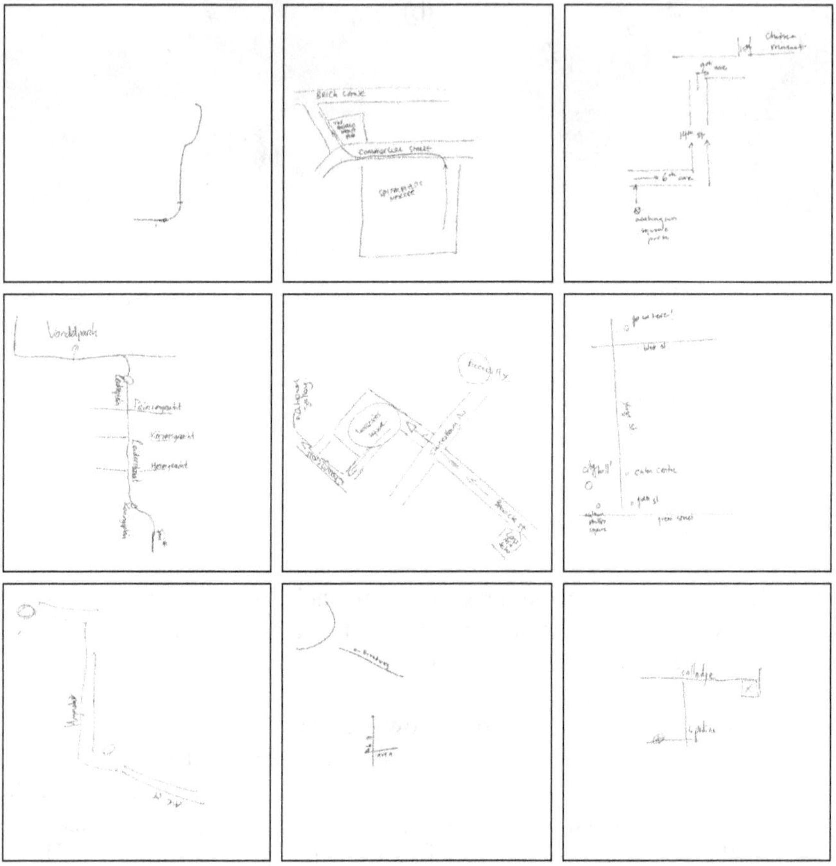

*Figure 4.2* A selection of directions from Amsterdam, London, New York City, and
        Toronto in 2017–2018.

of his observations, writes, "Obvious limits to such an undertaking: even when
my only goal is just to observe, I don't see what takes place a few meters from
me: I don't notice, for example, that cars are parking."[61] In other words, while
he produces legibility of a city, tracing and inventorying ordinary events that
may either go past unnoticed or animating what makes a city with the micro-
events that are its heartbeat. This is to make legible the everydayness of a
situation, cataloguing and indexing that which animates place. These can be
read as an augmentation of space through his experience of tuning into his
surround, a tracking down and a fleshing out an awareness of the "the infra-
ordinary."[62] The infra-ordinary fill in the creases and contours of the template.
    Perec and Brouwn's work is an analogue to the location awareness
animating locative media art. Locative media art brings together the affordances of mobility, interface, and location to consider a social and relational

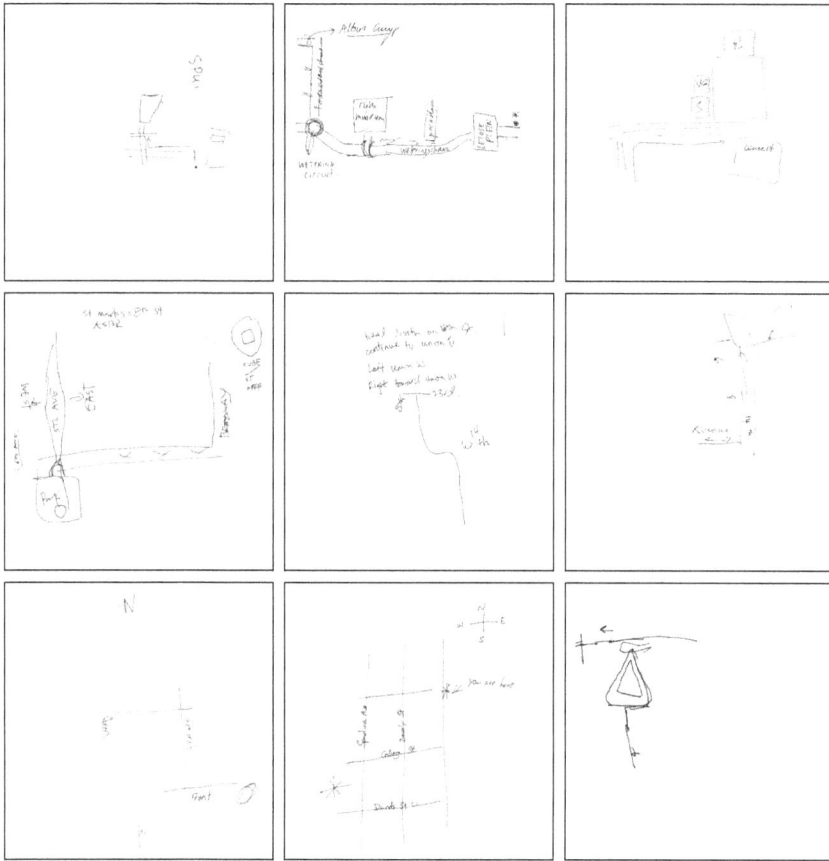

*Figure 4.3* A selection of directions from Amsterdam, London, New York City, and Toronto in 2017–2018.

aesthetics of location.[63] For example, artists like Janet Cardiff and George Bures Miller performed and recorded video walks such as "Night Walks for Edinburgh," using locative functions of GIS and GPS as well as the site-specificity of place, she used the map template to augment space, and to tell stories about it.[64] These are means to augment space. In this sense, AR art was its own kind of urban intervention, weaving in new readings of space such as *Clouding Green* by American artist Tamiko Thiel, commissioned in 2012 by the Zero1 Biennial in Silicon Valley, California.[65] The artwork is a large-scale augmented reality project that helps to visualise the 2012 Greenpeace report "How Clean is Your Cloud" that reveals the Green Energy Index of Silicon Valley's cloud computing providers. In the work, Thiel created AR visualisation of clouds situated over the headquarters of the big technology companies of Silicon Valley such as Apple HQ, Twitter HQ, Google HQ, Salesforce HQ,

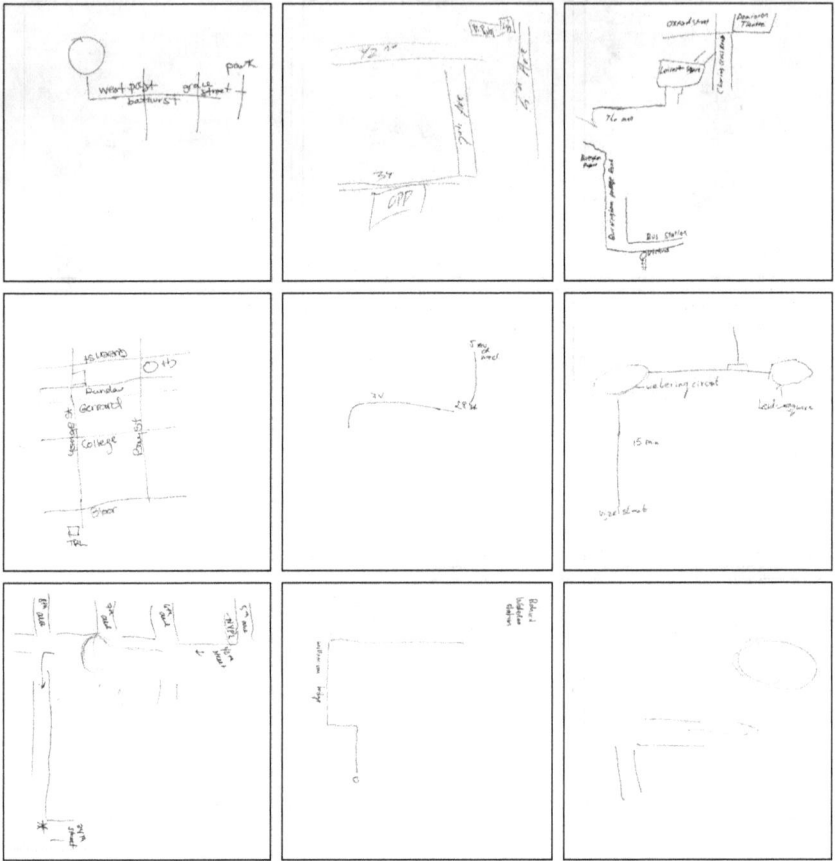

*Figure 4.4* A selection of directions from Amsterdam, London, New York City, and Toronto in 2017–2018.

and Facebook HQ. The clouds bear distinct colouration reflective of their carbon footprint, ranging from "sooty carbon black" to "brilliant renewable green."[66] For example, when over Salesforce, the cloud is a charcoal grey, signalling for the viewer a toxic grade on the Green Energy Index. Thiel plays with the materiality of the cloud and visualises the environmental impact of cloud storage. The project reveals the level of environmental impact these companies have on the environment, a material reality often obscured by the seeming immateriality of the cloud. The artwork opens up questions of how everyday information use intersects with the invisibilised impact of carbon emissions making legible how cloud companies have on-the-ground hazardous effects. Thiel's work brings awareness to the environmental toll of online information access, at a time when "going digital" and the internet is still positioned as a "green" choice.

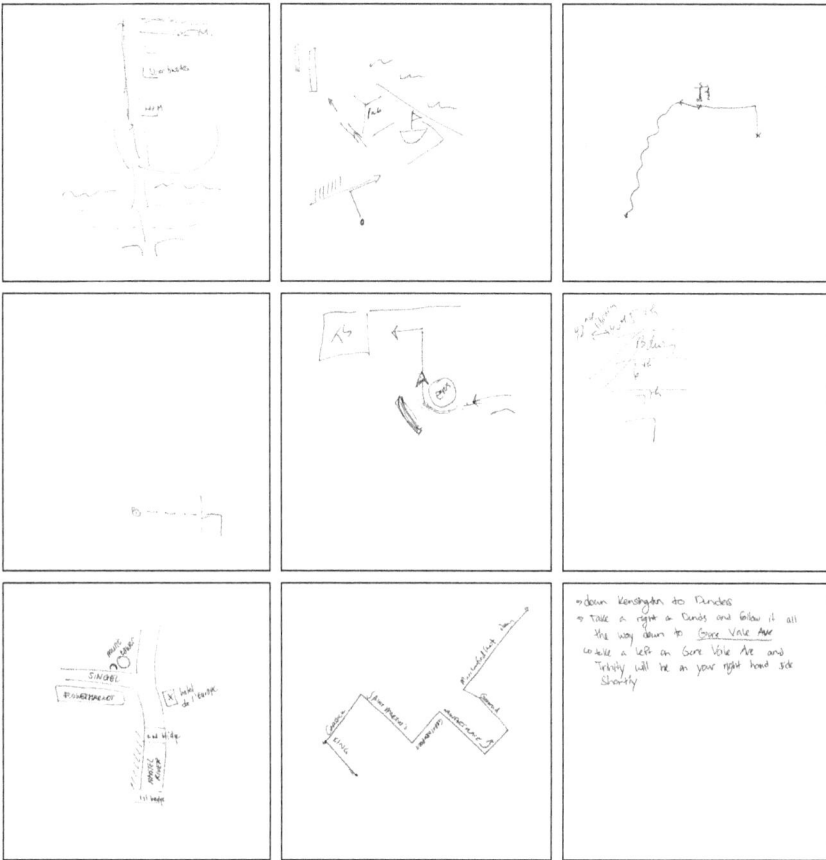

*Figure 4.5* A selection of directions from Amsterdam, London, New York City, and Toronto in 2017–2018.

In looking at AR art of StreetMuseum in London, to bring history to experiences of being in the city, Farman writes how these demonstrate an "emphatic experience" of the city.[67] The Museum of London's AR project StreetMuseum is layered with history that one can call up through the map. For Farman, the use of maps in locative media art calls up a "tension between the objective representation of space and the embodied perspectives on a space," which he asserts is "an important entry point into an analysis of the use of maps in locative media."[68] The question is, as Farman asks, is this experience "complimentary" to embodied experiences and does it relate to how one gets "their bearings" in space?[69] As Farman argues, these engagements of embodied experiences still perform the tension of prescribing what an embodied experience is in relation to. The question of what information is included in the representation of augmented reality remains, along with the question: what phenomenological response does the representation assume?

Recently, Google Maps has attempted to incorporate other phenomenological information of the city into its map. Google Maps' new initiatives for an "immersive view" of space expand on the map's legibility using augmented reality and 3D perspectives to simulate what an experience of space is. For example, Google Maps now incorporates augmented reality in its sharing of directions. Augmented reality directions featured on Google Maps function as a lens of urban legibility. In this feature, Google Maps is the overlay of "visual, immersive content on top of your real world."[70] By pointing one's phone camera at the street while using Google Maps' navigational function, animated arrows will appear on the screen indicating which direction to turn. In this sense, Google takes the guesswork out of deciding which direction to turn "according to the map" by appearing to look at the street and decide for you. These augmented graphics make the directions legible and help one feel more locationally aware (Figure 4.6).

Google's use of augmented reality incorporates another gesture into its form of location awareness—holding the phone up as if to take a photo. This gesture is meant to replace another common physical engagement with Google Maps' interface—one I encountered quite often in my experience of asking for directions—and that is the gesture of rotating the phone. When asking for directions, the gestures of looking at Google Maps and reading

*Figure 4.6* Illustration of Google Maps Live View in action. Illustration by Colin Medley.

space—while it might have been zooming in and out now it is something else. At times there was a disjuncture in relating the map's representation of space to the on-the-ground experience, and the result was a performative cartography of using the phone. This is most often manifested in the gestures of pinching the screen, but also rotating the map. For example, in New York, asking for directions to the New York Public Library, the person I asked told me that they did not know, so they looked it up on their phone, zooming in on the route directions to get the street names. However, even when using Google Maps, it became difficult to understand where we were, so they also rotated the phone in their hands to help orient us in relation to the map. The directions they finally gave ultimately led me south, away from the library, rather than north, towards the library.

In London, I was looking for directions to Trafalgar Square. The person first gave me spoken instructions on how to get there, and then, instead of drawing, they offered to *show me*, by looking the directions up on their phone. During this process of *showing me* the way, they kept rotating the phone to locate our position in relation to the map's positioning of us, even though the informant had already given me directions from memory. In New York, I approached people outside Penn Station for directions to Chelsea Market. One of them offered, "I'll look that up for you." They searched for the directions on Google Maps and then showed me the map. Rather than drawing the lines of the route directions, they wrote out directions based on Google Maps' text instructions, switching to the image to verify. During this process, they kept turning their phone around in their hands as they moved through the directions, in relation to the instructions they provided. In the end, they reassured me that the route I was to take was not very far, and that I only had to take "two turns," a number less than the number of times they rotated their phone in the act of giving directions. The Google affordance of location awareness did not seamlessly translate to the embodied sense of location awareness on the ground. The directions became difficult to translate from image to the ground. The gesture was not always helpful, nor did it make space any more clear.

Google Maps announced its Immersive View function with the promise of a "new generation of navigation."[71] Immersive View is the Google Maps 2D template rendered into a 3D city, combining the worldmaking affordances of online virtual worlds with augmented real-time feedback of location-based data. This combination of Google's aerial satellite images and street view images allows people to use Immersive View in reference to where one is or where one wants to go. London was part of the early prototyping of Google's Immersive View project. Google describes the project accordingly:

Let's say you're planning a trip to London. We can take you to Westminster and you can see where's Big Ben, and where's the London Eye, and where are they in relation to each other. We can then overlay additional information on top of this model of the world. If you want to see what it looks like with the sun coming from one direction or the sun coming from another direction, using all of this imagery that we've collected, we can actually put these experiences into what we've collected from Street View. And when you're ready to choose a restaurant, we can fly you down to street level and show you real-time busyness.[72]

Immersive View introduces other elements to spatial navigation like where the sun is at a particular time, the temperature, the sounds of the street, to then be translated to interior spaces of the restaurant to know how busy the place is. While many of these affordances are available on Google Maps' standard interface, Immersive View promises a new gaze tailored to the needs of the curious armchair traveller and the practical nightlife aficionado. The promise is that space becomes "intuitive," establishing what a city looks like in 3D and how a city feels. Therefore, Google Maps' Immersive View claims to make more than just directions through space legible, but also renders legible spaces' affects, like the vibe and the feel, or so is their promise.

Through Immersive View, Google contains the world to the terms of its legibility. Liz Reid, a VP of engineering at Google, describes Immersive View's feature as "massive zoom but on a neighborhood level."[73] Not only is it produced through Google's imagery—data generated from its own data collection, but Google's promise of emplacement annotated by tacit and embodied experience of space is aimed to replicate the dynamic experiences of being in place. These include details such as the height of buildings, the types of shadows a sun casts, the pedestrian traffic on the pavement. And these representations are in the name of containing experience itself offered to calibrate one's expectations of place, to anticipate how one might feel in that space. Immersive Views extends the template to a virtual reality and an affective experience of knowing a virtual world's promise.

With Immersive View, there is still the phone, the mediating interface, the environment on which space outside that environment is read. This interface, as Alexander Galloway has shown, is not outside of content but is always engaged in its own processes of display.[74] These are what Drucker pronounces as the "enunciations" of the interface. So while immersion suggests a dissolving of the interface to be fully in the environment, that is not, in fact, true. And indeed as Drucker argues, "graphical organisation only provides the basic structure of provocations for reading. The conditions constrain and order the possibilities of meaning production, but do not produce any effect automatically."[75] New forms are fixed to the screen but that does not change the flux of how they are interpreted or read. So while the graphical elements have changed, space is not necessarily any more legible. While the types of information are added to the interface and the dimensionality of that information may change, as well as how the screen layers with the world may change, it is still read through the screen. In this case, the screen contains, and like containers, promises to hold space in ways that are protective and modulated in the name of a smooth experience.[76]

But returning to the flux of performative cartography, reading Google Maps in relation to space can already be a form of immersion. This immersion takes the form of the dialogical relationship between representational objects like maps and our embodied engagement with space. In a letter to Leanne Betasamosake Simpson, Robyn Maynard describes using Google Maps Street View to tour downtown Toronto as a way of touring the violence of extraction and colonialism that appears placidly on the map. In moving through Google Maps, Robyn Maynard creates what she calls a "Google Maps itinerary" to "figure out how long it would take me to travel from my house to some of the places where our collective apocalypses are being drawn up."[77] Here Maynard is referring to Maynard calculating the amount of time it would take—one hour and twelve minutes to walk or twenty-eight minutes using public transit—to arrive at James Bay Resources Limited where the company, based in Toronto but extracting in the Niger Delta, drills for oil on Ogoni land. But the violence of this extraction is not detectable through Google Street View, obfuscating the decision made in the office. From there Maynard walks to Barrick Gold, mining around the world, notably in Tanzania, another site of "environmental and human atrocities."[78] From there Maynard walks to Belo Sun Mining Corporation, a company planning to build Brazil's largest open pit gold mine in the "still burning Amazon rainforest." The tour then ends with a look at the headquarters of Copper One, which once again, looks more like an administrative office than the home of ongoing extraction. Maynard concludes that "this tour of invisible carnage shows us that Toronto really is the global hub that it is so widely celebrated to be. It is on and around Bay Street that we find the direct lines between capitalist accumulation and those racial subjects whose lands and labour are *being* accumulated and poisoned."[79]

In keeping with the genre of Google Maps, Maynard includes the precise timing of the walks with the details of "9 minutes" and "8 minutes." The contrast between the precision of Google Maps' walking directions and the obfuscation or imprecision of what exactly is going on in those buildings serves as a metaphor for what is obscured and hidden—by the presence of a typical office building. They represent the geopolitics of space. Maynard narrates passing centres of colonial extraction and incarceration, that all present as benign office buildings, beige cement in downtown Toronto, but Maynard's tour is all about the violence that lies underneath. Maynard's tour offers a strategy of reading the map through its own metrics of legibility, in a world where colonial violence is naturalised. While a world away, each building is folded into the smooth functioning of the map, making the destruction they representbanal and ordinary in the organised interface.

Google encodes an imagined, embodied experience by adding details such as where the sun sits in the sky or how busy a place may be in that moment. In its attempts to translate an intuitive and immersive experience, it leaves out the fact that Google Maps is always mediating an immersive experience. Part of the tension of Google's immersion is that it smooths out what John Pickles terms the "silences of the maps."[80] Adding that these silences are not simply the missing, overlooked, or forgotten details but also the silent means by which the map works to overdetermine space.

What is unique about the digital and locative infrastructures is the immediacy to the person being located (i.e., the map is presented in direct relation to the individual using the map) and that it is encoded with "layers" of data which provide a semblance of complexity on the legible surface. However, Google's layers also "flatten" the city into new forms of totalising depictions that rationalise and tame the world. These homogenised abstractions of place erode the distinction of local differences and cultures in the name of producing legibility. Without the site specificity of place, space is made generalisable to, the words of Miwon Kwon, "better accommodate the expansion of capitalism via abstraction of space."[81] Legibiltiy can both forestall and foment a way of reading and being in space. Attempts to hold legibily, via templating or resurfacing only betray how slippery and confounding legibilty is. Moving from these orientations of legibility, the next chapter considers what it means to determine accuracy according to the terms of being read and the consequences and contingencies of presence and absence from the map.

## Notes

1  Google, "Explore and Navigate."
2  Google, "Explore and Navigate."
3  Kitchin and Dodge, *Code/Space.*
4  Mattern, *Deep Mapping the Media City;*
5  Mattern, *Code and Clay;* Mattern, *"Not a Computer."*
6  Google, "Our Approach."
7  Özkul, "The Algorithmic Fix," 594–608.
8  Özkul, "The Algorithmic Fix," 603.
9  Harvey, "The Spatial Fix," 1–12.
10  Özkul, "The Algorithmic Fix."
11  Verhoeff, *Mobile Screens.*
12  Verhoeff, *Mobile Screens,* 13.
13  Barnett, Prunty, and Rosenblum, "Handwriting Legibility Scale," 240–247.
14  Flesch, "Reply to Simplification," 54–55; Eleyan, Othman, and Eleyan, "Enhancing Software Comments Readability," 430.
15  Robinson et al., *Elements of Cartography,* 15; as quoted in Gaspar, Joaquim Alves, "Revisiting the Mercator World Map," 1195.
16  Bertin, *Semiology of Graphics.*
17  Bertin, *Semiology of Graphics,* 46. Emphasis is Bertin's, as quoted in MacEachren, "(re)Considering Bertin," 101.
18  Edney, "Some Thoughts on Jacques Bertin's."
19  November, Camacho-Hübner, and Latour, "Entering a Risky Territory," 582.
20  London Transport Museum, "Transforming the Tube Map."
21  Vertesi, "Mind the Gap," 7–33.
22  Transport for London, "Plan a Journey," results for a trip from Embankment Underground Station to Charing Cross on Tuesday, October 24, while showing the fastest routes using all transport modes.

23 Drucker, *Graphesis.*
24 Vertesi, "Mind the Gap," 7–33.
25 Vertesi, "Mind the Gap," 20.
26 Vertesi, "Mind the Gap," 9.
27 Vertesi, "Mind the Gap," 11.
28 November, Camacho-Hübner, and Latour, "Entering a Risky Territory," 582.
29 Oren, "Get Around and Explore."
30 Massey, *For Space*, 4.
31 Massey, *For Space,* 88.
32 Perkins, *Cartography: progress in tactile mapping*, Cole, "Tactile cartography in the digital age."
33 Frampton, Wong, and Solomon, *Cities without Ground.*
34 Dicker, "Street View Images."
35 Drucker, *Visualization and Interpretation,* 27.
36 Noone, "Locating Embodied Forms," 635–644.
37 Hesselberth et al., "Introduction," 7.
38 Gilmore, *Surface and Depth.*
39 Gilmore, *Surface and Depth*, 23.
40 Gilmore, *Surface and Depth*, 23.
41 Tuck and MacKenzie, *Place in Research.*
42 Gilmore, *Surface and Depth,* 27.
43 Mulvin, *Proxies.*
44 Smith, *Decolonizing Methodologies, 2nd. Ed.*
45 "Ogimaa Mikana: Reclaiming Renaming"; Lee, "Reclaiming/Renaming."
46 "Ogimaa Mikana: Reclaiming Renaming."
47 Sharma, "It Changes Space and Time! Introducing Power-Chronography.
48 Smith, *Decolonizing Methodologies.*
49 Harvey, *Paris, Capital of Modernity.*
50 do Couto, "Ideology and Utopia," 730–731. Rüegg (ed.), *Brasilia.*
51 Smith, *Decolonizing Methodologies, 2nd Ed.*
52 Lynch, *Image of the City.*
53 Lynch, *Image of the City,* 128.
54 Lynch, *Image of the City,* 133.
55 Lynch, *Image of the City,* 134.
56 Lehmann, "Stanley Brouwn," 55.
57 Brouwn, 1964, cited in Russeth, "Stanley Brouwn."
58 Lehmann, "Stanley Brouwn," 60.
59 Perec, *An Attempt at Exhausting*; Licoppe, "Georges Perec, Observer-Writer"; Highmore, "Georges Perec."
60 Highmore, "Georges Perec," 106.
61 Perec, *An Attempt at Exhausting,* 10; Highmore, "Georges Perec," 106.
62 Perec, "Approaches to What?", 206; also quoted in Livesey, "From the Infraordinary," 2.
63 Wilken and Goggin, *Locative Media.*
64 Cardiff and Miller, *Night Walk for Edinburgh.*
65 Thiel, "Clouding Green."
66 Thiel, "Clouding Green."
67 Farman, "Map Interfaces."
68 Farman, "Map Interfaces," 86.
69 Farman, "Map Interfaces," 87.
70 Daniel, "Immersive View Coming Soon."
71 Daniel, "Immersive View Coming Soon."
72 Daniel, "Immersive View Coming Soon"; Dicker, "Street View Images."

73 Pierce, "New 'Immersive' View," quoting Liz Reid.
74 Galloway, *The Interface Effect*.
75 Drucker, *Visualization and Interpretation* 102.
76 Sofia, "Container Technologies," 181–201.
77 Maynard and Simpson, *Rehearsals for Living*, 11.
78 Maynard and Simpson, *Rehearsals for Living*, 11.
79 Maynard and Simpson, *Rehearsals for Living*, 11.
80 Pickles, *A History of Spaces*.
81 Kwon, *One Place After Another*, 158.

# 5 Orientations of Error

**Faults | Absences**

Google Maps errors are something of legend. The stories of Google Maps-related mishaps recount cars driving into swamps or hikers rerouted through treacherous mountain trails, conveying both surreal misadventure and cautionary lore.[1] Some incidents make news headlines like "Google Maps Leads About 100 Drivers into 'a Muddy Mess' in Colorado"[2] and "Google Maps confuses N.J. House for State Park, homeowners are besieged by 'belligerent' visitors."[3] Over the years, reporters have chronicled a variety of Google Maps direction blunders, from pedestrians walking along the narrow shoulder of a busy highway,[4] to hikers stranded in the Grand Canyon. Some accounts relay the consequences of Google's misdirections. For example, in 2016 Newsweek reported on a demolition team levelling the wrong home in Texas thanks to Google Maps' incorrect address listing.[5] Or there is the 2010 CNN report about Google Maps mislabelling the Costa Rica-Nicaragua border, where, according to reporter John Sutter, "Nicaraguan troops crossed over, removed the Costa Rica flag and placed down their own" based on Google Maps' spatial projection on the disputed territory.[6] While the stakes of each incident vary, these testimonies are at once morality tales (they should know better), advisory messages (pay attention!), and a titillating piece of gossip (can you believe it?).

While the headlines might suggest Google Maps is to blame for these mistakes, the active agent in these errors, the stories also invite judgment on the user for *making* the error. Surely one cannot expect a viable route to an airport through a bog? Technology journalist Jacob Kastrenakes epitomises this sentiment in his article titled "Do not blame Google Maps when you tear down the wrong house"[7] in which he recounts in disbelief that, "rather than shouldering the fault for this error, someone from the demolition company apparently pointed to screenshots of Google Maps." And while demolishing the wrong home is an extreme example, Kastrenakes articulates a popular imaginary of what a Google Maps "error" can be in the landscape of location awareness: a simple mistake activated and aggregated by individual failure or at least a deficit of skill.[8] As I trace throughout this book, a sense of direction is not "common" or shared, nor is Google Maps' brand of location

DOI: 10.4324/9781003251569-5

awareness a tacit "common sense." Instead, location awareness plays out in the folds of everyday experiences of space. Similarly, navigating different interpretations of space, however divergent, enacts a sense of direction. The question is, what relations to space do Google's errors condition beyond simply following the *wrong direction*?

In Chapter 4 I argue Google Maps claims to make space legible by keeping out the disorienting chaos of any given environment. These terms of legibility are echoed in its quest to *add* more and more information to the map in the name of building out the accuracy of its map. But adding information occurs according to the rules and contours of the maps' legibility in the name of maintaining the stability of the representation. All that does not fit will either be left off the map or if it is to be included, it will be included according to the maps' terms. A preoccupation with accuracy and error is often a question of absence and presence—a missing turn, a void in instruction, a new feature in Google's instruction, or a immersive view. Google's conditioning of error doesn't just inform Google Maps' territory but informs the conditions of using the map and belonging on the map. When we think about Google error in these narrow terms, these bounded routes of accuracy, the question is: what other kinds of information swamps can we get stuck in?

This chapter is about errors revealed through absence and the ways Google subsequently manages such absence. Sometimes absence is a precise omission passed off as a simple mistake, as I will explore in Buffalo, New York's Fruit Belt neighbourhood. In other ways, absence becomes opportunity for participatory strategies of adding data and the fault lines those create in the map, as I explore with Google's now shut down MapMaker tool. Finally, I look at where addressing absence is leveraged as inclusion where people are made *to be* precise coordinates of the map, as is the case with Google's Plus Codes project. Here Google's orientation to error ascribes absence in the name of accuracy. Error defined through absence introduces a double bind of inclusion over exclusion, using inclusion as a method of calibration to the map's organisation of space. Google Maps' location awareness is oriented not just by space, or local relations to space, but also by the proprietary goals of Google Maps' global mapping project. The result of who and what gets mapped in the name of *a right way* is more than a question of direction, or the question of location, but a question of who and what does Google Maps invest in when it represents space, and what becomes an error by way of simple fault or absence.

**A Fault in the Map**

Accuracy, reliability, and completeness are Google Maps' implicit and explicit promises. From projects for collecting spatial data like Ground Truth to the affordance of Google's real-time traffic updates, Google Maps sells accuracy through claims to "innovations and investments that help build an accurate and up-to-date understanding of place for the billions of people looking for local information on Search and Maps."[9] In Chapter 2, I discuss Google's construction of objectivity. Here I discuss accuracy as the residue of that objectivity. Google's promise of accuracy is one of the mechanisms through which it exercises its authority, and Google's Terms of Service and the Google Maps/Google Earth Additional Terms of Service are the contracts that delineate the limits and margins of accuracy. The third clause of the Google Maps Terms of Service, titled "actual conditions; assumption of risk," declares that when "you use Google Maps/Google Earth's map data, traffic, directions, and other content, you may find that actual conditions differ from the map results and content, so exercise your independent judgment and use Google Maps/Google Earth at your own risk. You're responsible always for your conduct and its consequences."[10] The language of risk acknowledges an inevitability of error (that is why there must be an agreement) as a condition of Google Maps' world map. The imperative of independent judgement overdetermines the individual as responsible for maintaining the stability of the map, thus absorbing the blame for all errors. In Google Maps' terms, accuracy is a legal construct, oriented towards authority without liability and consequence.

More than a contractual agreement, Google's claims to accurately mapping the world rely on a series of tacit social agreements and constructions. For example, according to Luiz André Barroso, VP Engineer at Google Maps (remember him from Google's "Why We Map the World" video from Chapter 3), the great challenge to Google Maps' grand design is the "dynamic" world where data can quickly become "stale."[11] Barroso's claims discursively lay out the terrain of the Google Maps imagination: a grand narrative that *tames and enframes* the unruly world flexible to updates and renewals.[12] Rendering the world as "dynamic" is a rhetorical technique that flattens the geographical, linguistic, and the sociopolitical configurations of "difference." The term "dynamic" here is doing a lot of work of evening out difference, making unevenness appear inevitable. Similarly, "stale" portrays information as that which must be simply refreshed as a form of intervention into the cobwebbed and dusty corners of Google Maps—add in a few new road signs, new storefronts, and perhaps some new zoning. Barroso's dynamic-stale binary reduces space to a computational formula through which accuracy is achieved through procedures like the terms of service.[13] Yet the terms by which information is dynamic are never explicitly defined. Instead, accuracy is leveraged as a means of taming the dynamic world through the stability of precision.

In Barroso's metaphor, the collision of stale and dynamic data create friction, or a glitch. Perhaps when we think of the glitch we think of a frozen screen, a pixelated image, or a fissured sound—but a glitch on Google Maps is more than simply a temporary disruption of otherwise seamless location awareness; glitches expose the assumptions about what fits into the map in the first place. Ruha Benjamin argues that glitches are indeed social processes used to regulate "normative world building." In her analysis of a 2013 tweet from media specialist Allison Bland,[14] Benjamin reflects on how "default discrimination" embedded in technologies like Google Maps often present as a simple glitch. Bland's tweet from 2013 reads, "Then Google Maps was like, 'turn right on Malcolm ten Boulevard' and I knew there were no Black engineers working there." Through this tweet, Bland recounts her experience of using Google Maps' voice navigation system. The audible translation of Google Maps' turn-by-turn directions, being told to "turn right on Malcolm ten Boulevard,"[15] misinterprets the X in Malcolm X to be the Roman numeral ten instead of the name of a Black Liberation leader. Both Benjamin and Bland point to how being told to "turn right on Malcolm ten Boulevard" is not a simple misinterpretation of the X in Malcolm X.[16] This spatial directive also erases a historical figure, displaces the significance of the "X" in Malcolm X, and recodes the territory where Malcolm X Boulevard is situated. The naming of Malcolm X Boulevard reflects a movement of renaming in which, as Zaheer Ali describes, "Malcolm X has become a place onto its own."[17]

In her tweet, Bland points to the absence of Black programmers at Google Maps as a possible reason for this misclassification of "X." This absence is indicative of a social process and not just a simple mistake. Adding to absence, Benjamin argues that "the Google Maps glitch is better understood as a form of displacement or digital gentrification mirroring the widespread dislocation underway in urban areas across the United States." Bland and Benjamin demonstrate how this glitch in the map is actually "systematic, structural, routine." The glitch as such is a social process that masks default discriminatory designs as if "a fleeting interruption of an otherwise benign system, not an enduring and constitutive feature of social life."[18] The Google Maps glitch reifies racist spatialities and reinforces ongoing processes of displacement. Therefore, it is more than simply that Google Maps doesn't process race but instead Benjamin and Bland's analysis of this experience reveals race and class are embedded in Google Maps' design processes and classification systems. This misclassification of "X" reveals a pervasive whiteness of the digital world, from design to circulation, and more specifically how casually racist ontologies inform Google Maps' conditions of accurate navigation.

Drawing from Legacy Russel's *Glitch Feminism*, the glitch can be an important discursive manoeuvre for both obscuring and maintaining power. On one end, the glitch is what Russel describes as "an indicator of something having gone wrong."[19] Russel notes the glitch marks an inversion—a rupture, a stifling, or a failure to compute—they also shift the understanding of what the *error* of the glitch is—the system. Glitch is the dissonance of not fitting in, not meeting expectations, or what Russel frames as a deliberate "non-performance" of fixed hegemonic norms.[20]

When the glitch presents as a technical process then the solution to the glitch is also technical: a reboot, a shutdown. But the easy technical fix reinforces what Lauren Berlant describes as "the repair of what wasn't working."[21] Berlant continues, "in the episode of a hiccup, the erasure of the symptom doesn't prove that the problem of metabolizing has been resolved; likewise, the reinitialising of a system that has been stalled by a glitch might involve local patching or debugging (or forgetting, if the glitch is fantasmatic), while not generating a more robust or resourceful apparatus."[22] Google mobilises the imaginary of the simple glitch to orient mapping errors as temporary and technical. Seen as temporary and infrequent, the glitch is "no big deal" or "just a momentary technical issue" it reassures—"don't worry, we'll be back on track soon." Such errors on Google Maps are more than temporary technical mistakes but formed from mapping's enduring legacies of assigning fixity.

To reorient the error as a condition of fault or absence, I turn to Google's mapping of Buffalo, New York. In 2008, in the early years of popular mobile mapping media, Buffalo's historic Fruit Belt neighbourhood was missing from Google's mapping of the area. In its place was the neighbourhood label of Medical Park.[23] According to Caitlin Dewey's reporting, the Fruit Belt's absence was first discovered by Fruit Belt resident Veronica Hemphill-Nichols. Hemphill-Nichols who Googled directions to her house on her smartphone, which in 2008 was still a novel experience given Google's launch of its mobile application around this time. When the map loaded, she saw the directions led her to Medical Park and not to the Fruit Belt.[24] The Fruit Belt was nowhere to be found—struck from the spatial record. In 2019, the issue came back to life as, making news headlines once again, a reality that suggests a lingering threat of a name change.

At first, this might seem like a simple misnaming. Some crossed wires, some missing inputs. A quick fix to a momentary glitch. But looking closer at Google Maps, the history of this neighbourhood, and the maps that have surveyed these neighbourhoods over the years, the renaming of the Fruit Belt is anything but a one-time incident. Instead, Google Maps' indexing of space is reflects the ongoing expansion of Medical Park into the Fruit Belt.

The Fruit Belt neighbourhood is a majority-Black residential neighbourhood in New York State's second-most populous city—Buffalo. It is a neighbourhood, with a track record of strong local organising practices that "fight against gentrification and systematic erasure."[25] The Fruit Belt Community Land Trust, a grassroots organising effort, describes their neighbourhood's history as "seventy plus years of African American community building in the face of injustice."[26] The Community Land Trust identifies this injustice as "systematic and systemic racism demonstrated through redlining, loan denials, mortgage discrimination, restrictive covenants, block busting, and urban removal masked as 'Urban Renewal.'"[27] Ongoing efforts of systematic erasure included the City of Buffalo's demolition of the Ellicott District, "leaving 1,900 mainly African American families homeless."[28] Throughout its history, the Fruit Belt has resisted and refused the targeted and persistent racist policies and practices premised on erasure and dispossession.[29]

The marker "Medical Park," on the other hand, is the name given to a series of new commercial and residential development projects encroaching on the Fruit Belt neighbourhood. Hemphill-Nichols and other residents have long been organising against the Medical Park's development "boom" and its reverberations into the Fruit Belt.[30] Today, Medical Park is also known as the "Innovation District," home to the Buffalo Niagara Medical Campus, Inc., a 120-acre medical campus that comprises industrial-scale building projects such as hospitals, training facilities, and research labs. The campus is nestled in area of rapid rebuilding including the construction of new multiunit residential complexes and spaces to welcome expanding commercial enterprises looking to move to Western New York. The Medical Park development discourse touts the promise of "building a dynamic entrepreneurial ecosystem" based on "attracting investment to Buffalo."[31] The Medical Park's housing initiatives target doctors, medical students, and biotech entrepreneurs to attract people to the area rather than support the people who are already living there. More than simply a hospital, Medical Park is a commercially minded environment augmented by residential and commercial development further encroaching on the areas that are built as new social infrastructures of so-called innovation.

The Medical Park's expansion into the Fruit Belt is the latest iteration of predatory land claims that these Buffalo residents have faced over the years. For example, the Kensington Expressway, built in the 1970s, routed directly through the Fruit Belt in the name of connecting more people to the newly expanded Buffalo General and Roswell Park Cancer Institute.[32] According to the Fruit Belt Community Land Trust, the building of the expressway and subsequent expansions of the medical campus demolished nearly 40% of housing in the Fruit Belt until a moratorium on demolitions was enacted in the early 2000s. The Medical Park's encroaching proximity to the Fruit Belt is no accident but is part of histories of dispossession. The ongoing survival and thriving of the Fruit Belt community shows that it is no stranger to forms of racist removal and emptying of land to be commercialised, upsold, and gentrified. Google Maps' renaming practice is consistent with the racist antagonism threaded throughout the Fruit Belt's history.

According to Caitlin Dewey, Google Maps' renaming of the Fruit Belt neighbourhood was a decision based on old city data along with "defunct mapping start-ups, and a ubiquitous, secretive data broker that claims to keep tabs on 100,000 neighborhoods."[33] While the origin or data source speaks to Google's opacity of data sourcing, it also points to the structural data issues Google Maps is imbricated in. The secretive data brokers or the transient mapping start-ups illustrate the profit-driven landscape of data mining and data selling. Locational data, like real estate, is a speculative and profitable game of investment.[34] Therefore, it is not a question of whether Google Maps created a *new* name for the Fruit Belt neighbourhood but instead, it is a matter that Google Maps replaced Fruit Belt's name with Medical Park, ultimately choosing which data is more profitable. Google Maps chose a side in this ongoing struggle of land rights and affordable housing—the side with money—ignoring the documented history and current organising efforts of the Fruit Belt.

The renaming of the Fruit Belt is a form of neighbourhood rebranding—an action in the name of service to the real estate market. According to an interview with Hemphill-Nichols in 2019, the renaming empowers developers to continue building at the expense of affordable family housing. And while Buffalo Niagara Medical Campus positions itself as a non-profit institution, their priorities are commercial investment, innovation, and economic opportunities.[35] The Medical Park's investment promises are based on real estate speculation and development. And while this is not to attack health care specifically, but to look at city-sanctioned and institutional forms of racism based on the dispossession of racialised and classed Americans. Hemphill-Nichols identifies how Google Maps' role in labelling neighbourhoods enforces a takeover when she states, "they took (the name) from us" and "we want it back." Her words demonstrate how Google Maps' labelling legitimises the presence of the Medical Park developments and the white affluent classes it ultimately serves.[36] Google Maps' renaming is a valuing of land in the name of entrepreneurship, innovation, and development and the devaluing and exclusion of Black livelihoods. In 2019, 11 years after she noticed the naming error, Hemphill-Nichols is still fighting—a preoccupation that points to the precarity of the Fruit Belt on Google Maps.

The prominence of the Medical Park label on Google Maps might seem like a simple error—as if Google simply used some "bad" data or *forgot* to add the Fruit Belt. Under these conditions of error, such a "glitch" is something noticed and then corrected. But the timeline of this issue—first reported by Hemphill-Nichols in 2008 and then returning as an issue in the press headlines in 2019—shows this renaming is an ongoing story. Like gentrification and encroaching development, it does not simply happen one time. It's ongoing—a replacement tactic, like the renaming of Malcolm X Boulevard that marks what is "of interest" for Google Maps in the ongoing practices of land speculation in Buffalo. Google Maps is not a neutral player in this fight. Indeed, the context reveals Google's conditions of accuracy to be based on a set of priorities undergirded by racial capitalism and speculative real estate.[37] The Medical Park side is aligned with the development of condominiums, private investment, and higher real estate value. Google Maps chose a side, and continues to choose a side, as the inclusion of the Fruit Belt on the map seems to be always under threat, from 2008 to when organising gained media attention in 2019. Google's indexing of space and the priorities of this index are in lockstep with what Katherine McKittrick calls "spatial colonisation and domination: the profitable erasure and objectification of subaltern subjectivities, stories, and lands."[38] In other words, this is not a glitch—or a mistake that can be responded to and then quickly resolved but is instead evidence of Google Maps' fundamental priorities of making money off racist systems of displacement.

Absence from the map is consistent with development and planning projects taking place in Buffalo. According to Dewey, other Buffalo neighbourhoods such as Kensington-Bailey, Allentown, or the Elmwood Village have reported similar conditions.[39] Google's mapping of the Fruit

Belt is the reinforcement of segregationist planning policies and the undermining of Black resistance. Like the erasure of Malcolm X, the erasure of Black lives from Google Maps is not a glitch but consistent with what McKittrick terms "cartographies of struggle" in which "concealment, marginalization, boundaries are important social processes."[40] Google's renaming of the Fruit Belt illustrates how Google Maps can be a tool of dispossession and displacement when it operates through and alongside processes of racialised expropriation.[41]

As more than a simple error by way of absence, the expurgation of the Fruit Belt is not a *new* racist data enclosure, but an extension of American redlining practices discussed in Chapter 2—the same practices the Fruit Belt community have been challenging and resisting for decades. More specifically, Google Maps' erasure of neighbourhoods like the Fruit Belt is a process of *digital* redlining.[42] Google Maps continues the logic of redline mapping, deciding who and what has value and who and what is a threat to value.[43] As Aaron Shapiro has argued, such location-based information "abstracts place by datafication," making algorithmic linkages that prescribe value on space, as places to be in or places to visit or neighbourhoods to buy property in.[44]

Following Sara Safransky's work on redlining in smart city infrastructures, data-driven violence in the digital realm further entrenches racism and inequality.[45] Google's representations circulate as "objective truth" about the world through its API. In other words, the value of whiteness embedded in Google Maps' organising logic permeates far beyond the map itself. Therefore, it is imperative to see Google's erasure of the Fruit Belt as not just a glitch that reinforces oppressive social relationships and enacts new modes of racial profiling. These are new modes of racial profiling with what Lisa Nakamura identifies as racist classification schemes baked into the system.[46] Yet, this coding of space is rarely labelled "misinformation" or a system in need of a comprehensive edit and moderation. Instead, these inequitable classification systems inform what is presented as accurate forms of location awareness on the map, serving to, as Safiya Noble contends, *mask and deepen inequalities.*[47]

Google's plotting of "Areas of Interest" or AOI,[48] a term we will return to later in this chapter, reflects a form of venture mapping, or investment in business-related and consumable space. AOIs are visible on the map in yellow/orange shading to mark the commercial corridors.[49] According to Justin O'Beirne, Google generate AOI's from Google Maps' own data, defined through frequency of visits, satellite imagery, and street view numbering.[50] Google's tracking of such areas betrays its investment in commercial interest and its priorities in its platforming of space. AOI's are a process of over-determining which neighbourhoods are of interest to people who are, as Sarah Sharma argues, "being located into even more normalizing and knowable mobile populations."[51] Accordingly, Google Maps weighs in on the ongoing disputes between the Fruit Belt and the Medical Park areas of New York. Google Maps operates through a coding of commercial interests and while that is convenient in the case of looking up a restaurant, this coding has serious repercussions for how space is organised, how value is distributed, and how information about space circulates.

Importantly, the Fruit Belt's resistance to Google's indexing of space demonstrates a legacy of displacement that has long been challenged, echoing McKittrick's question of what if "we insist that past and present geographies are connective sites of struggle, which have always called into question the very appearance of safely secure and unwavering locations?"[52] The Fruit Belt residents are not just the "actual conditions" marked in Google's Terms of Use, but they are also refusing Google's imposed risk of erasure. The Fruit Belt's organising efforts reorient an understanding of absences from the map as more than a simple error. While Google's brand of data capitalism may suggest "if you are not on our map, you do not exist" as their cartographic truth, drawing attention towards the buzz of the glitch, the short-circuiting of a connection, can be an act of resistance that draws attention and rejects hegemonic renderings of space.[53] Even when oriented as an error, the glitch can call out.[54] They show us that Google's orientations are not inevitable, and people will not be duped. That, to paraphrase, bell hooks, while the map might create its own spatial creases and margins, these are not empty spaces but are themselves sites of resistance.[55] While Google may attempt to use the glitch as a strategy to abdicate responsibility, we can see how it extends the indexical ontologies of racial capitalism where the redline and the rename enact similar functions.

Beyond Buffalo, Google Maps often adds new real-estate value-driven names to neighbourhoods. These include names like Vinegar Hill Heights and Midtown South Central appearing in New York.[56] When I open Google Maps in the city of Toronto where I am from (but no longer live), I am perplexed by the labelling of neighbourhoods such as West Bend—a name I have never heard even though I grew up in the area Google marks as West Bend. This defamiliarising of the familar parallels the ongoing developments of contested high-rise and high-value condominiums' and townhouse projects' rebranding of the area. In August 2018, Jack Nicas at the New York Times reported new naming protocols in San Francisco and Detroit that were "given life" by Google Maps. In San Francisco, the neighbourhoods of Rincon Hill and South Beach, located just south of San Francisco's downtown, were recently "rebranded" on Google as "The East Cut."[57] The rebranded name is not the creation of Google, but the product of the area's business improvement association (BIA), predominantly represented by large technology firms such as Salesforce, an internet marketing analytics company. According to Nicas' reporting, the BIA elected to change the name of the area to The East Cut without public input. Once the BIA decided on the change, they notified Google Maps to rename the area. The name change went largely unnoticed until changed on Google Maps and proliferated through the other digital maps that use the information. These practices of renaming are akin to the rebranding in the Fruit Belt where Google works as a megaphone for commercial interests, reinforcing the social conditions of locational value.

At the same time, Google Maps extracts value from all that it maps. As previously discussed in Chapter 2, the Google Maps API circulates these neighhourhood names beyond the Google Maps interface as a form of intellectual property. They become embedded placenames in other location-based platforms like Zillow or Uber.[58] This information is then presented as truth on other platforms or used in other systems specifically like Zillow that evaluate property value in the name of facilitating the sale of real estate. Google Maps operates as a form of digital redlining. While Google purports a commitment to accuracy, its indexing of space mobilises and profits from existing systems of discrimination and exclusion. Google Maps' investment in digital gentrification results in displacement and dislocation not just from the map but from the places that are mapped.[59]

While the digital maps may show us new forms of circulating unjust data, the politics of geographic information remain relatively consistent.[60] Google Maps is consistent with a pattern of digital redlining where, as Gilliard reflects, "personalisation and monetisation works perfectly well for particular constituencies but it doesn't work quite as well for persons of colour, lower-income students, and people who have been walled off from information or opportunities because of the way we are categorised according to opaque algorithms."[61] Beyond profiting from data, Google Maps is making a choice to actively reinforce boundaries of race and class creating the digital blueprint for data regimes beyond Google Maps. Looking closer at Google Maps and its mapping practices we notice how the Google Maps glitch reflects André Brock's critique of the Internet and how a "refusal to mark whiteness as an identity powers the concept that internet culture is raceless."[62] Racism is more than just temporary, it is enduring, structural, and systemic. On Google Maps, this structural racism informs what data appears on the map, what data is shared through the map, what is omitted, and what relations to space are valuable.

**The Shutdown**

What happens when the so-called glitch reveals the logic of error and accuracy in Google Maps in a way that is unaffordable to Google? In other words, what happens when the glitch reveals a pattern, rather than an isolated incident, and this pattern translates to bad press? The Google Maps tactic of controlling the endemic glitch is to shut down the public-facing mechanisms of production, as evidenced through its handling of its Map Maker Tool. Since 2008, Map Maker has been the crowdsourcing arm of Google's Ground Truth Project.[63] Google's Map Maker Tool enabled "users" to "share" their local knowledge in the name of "improving the map" but without programmatic access to Google Maps itself.[64] In other words, changes suggested through Map Maker were then formally added to Google Maps and this information became Google's property. This was clearly stated in the Google Maps Terms of Service which gave Google Maps "perpetual, irrevocable, worldwide, royalty-free, and non-exclusive license to reproduce, adapt, modify, translate, publicly perform, publicly display, distribute, and create derivative works of the User Submission."[65]

With Map Maker, Google was able to access detailed information and data provided by a global network of people who used Google Maps. After the launch of project Ground Truth, Map Maker was especially useful for Google Maps since there were so many areas around the world Google Maps had yet to access with its Street View vehicles, satellites, and government surveys. For Google Maps' grand design of the universal map, this was a problem. With Map Maker, ordinary people with access to the internet and with some degree of technical proficiency could add map details like roads, businesses, or traffic lights. Map Maker helped Google claim an accurate spatial projection. Anything that was missing or out-of-date could be flagged by an on-the-ground informant, someone with knowledge about the "actual conditions" (Figure 5.1).

However, this crowdsourcing sometimes introduced problematic errors into the map. Google has historically addressed crowdsourcing errors through the framework of "bad actors" who "try to mislead people through a variety of techniques, from fake reviews that attack a business to inauthentic ratings that boost a place's reputation."[66] In 2015, Map Maker's so-called bad actors became an overwhelming problem for Google Maps. First, there were reports of Google Maps' racist search results when the input of a racist epithet paired with the word "president" directed Google Maps users to the White House during the Obama presidency.[67] This incident followed the misogynist re-naming of women's dormitories at a school in India.[68] The final straw for Google came after reports that when zooming into the map in the territory outside Rawalpindi, Pakistan one was greeted

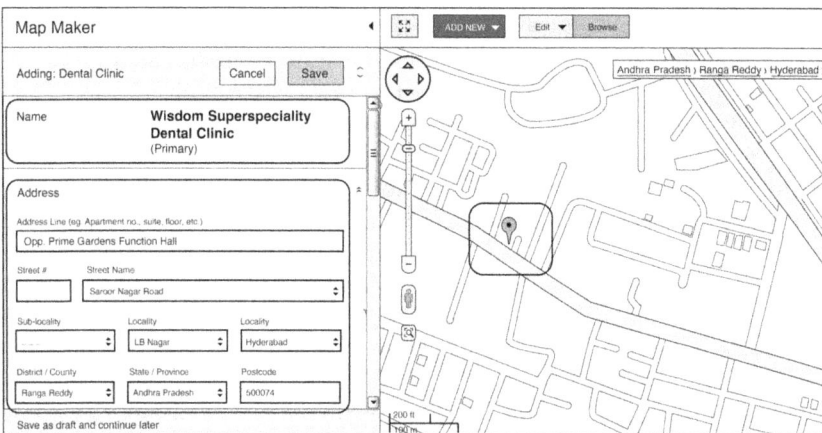

*Figure 5.1* Illustration of Google's now-defunct Map Maker Tool. Illustration by Colin Medley.

with an image of an Android robot urinating on the Apple logo captioned "Google Review Policy is Crap."[69] Commentary on these incidents in news outlets and tech journals framed the racist renaming "hackings" and "vandalism" situating these actions as a defacing of map property at the hands of some "bad actors." The labelling of "bad actors" is a distancing tactic that places wrongful data as the fault of outsiders in Google's imagined crowdsourcing community. These outsiders cause the glitches that sabotage mapping efforts.

Following these incidents, Google enlisted the "nuclear option"—in other words, an emergency shutdown of the entire Map Maker system. Pavithra Kanakarajan, Google Map Maker product manager, blamed the shutdown on a backlog of edits as the company moved towards "a manual review process" for all edits. Kanakarajan framed the discontinued operation of Map Maker as "not fair to any of our users to let them continue to spend time editing. Every edit you make is essentially going to a backlog that is growing very fast."[70] These new protocols led to slower input processes, effectively overwhelming the system, demonstrating the precarity of Google's review process and bringing into focus how faulty Google's review process is. Google once again points its finger at the "bad actors" that have overrun the system. The threat remains external to Google Maps rather than constituted through a position that a singular, total, universalising map can be possible, and is an ambition that evades politics. Map Maker illustrates the failure of the one-size-fits-all approach to mapping moderation and mitigating online harassment.

The shutdown, or the "nuclear option," is another strategy of containment, this time rather than minimising or making fleeting, the shutdown disappears the error, in a "nothing to see here" deflection of the mistake. Google does not shut down the entire Google Map platform, of course. Instead, it locates and isolates specific mapping tools as the source of the virulent glitches, and effectively cuts the power. In this case, the "open source" Map Maker Tool. Like the glitch it tries to conceal, "the shutdown" is a means by which Google Maps orients error as the fault of a few "bad apples" rather than a constitutive feature of the map itself.[71]

Despite this global shutdown, Google continues to celebrate Map Maker as a hallmark of digital mapping innovation. In 2021, on the 15th anniversary of Google Maps, Google CEO Sundar Pichai reflected on the history of Map Maker as "a way for people to add streets and local landmarks to improve the experience of Google Maps." For Pichai, Map Maker was a central feature of Google Maps' global development and, moreover, a key indicator of Google Maps' social benefit. In celebration of Map Maker, Pichai declares "not only do maps make it easier to get around; they also can give you a sense of identity when you identify your street on the map for the first time. That was one of the revelations of Map Maker." According to Pichai, Map Maker "revealed" to Google the power of people adding their street to the map as a form of cultural identity and as a way to situate oneself inside of a global system. Map Maker's social benefits solidified through how it is used in the everyday heroics of mapping noting how Map Maker "quickly evolved to help map floods in the Philippines and Pakistan, and later to allow people in the U.S. add a new road to their neighborhood."[72] In his speech, Pichai revises the memory of Map Maker as a tool that made the world better rather than reinforced and amplified the discriminatory systems that are, indeed, spatial.[73] The events that led to the shutdown are a forgettable glitch.

Shutdown becomes the method of correcting Google Maps' errors, but the process of shutdown also abdicates responsibility. As Lisa Nakamura and Charleton McIlwain point out in their Wired Magazine interview following the shutdown, Google's shutting down the Map Maker Tool does not erase (nor does it even effectively contend with) the realities of racism and inequality online.[74] Instead, it closes off one area where it is overt whilst actively perpetuating racism when it's embedded. While the glitch is an attempt to deflect the responsibility and obscure ongoing discriminatory practices that undergird Google's mapping project—or make illegible through the map—the total shutdown becomes the option when we have no choice but to reckon with these errors. The shutdown action makes it seem like the Map Maker systems (and Google Maps) are in a state of emergency, a melting down, rather than functioning as it always has. This is not to say that racism, discrimination, and misogyny are inevitable in themselves, but that online harassment exposes the map's participatory infrastructure as a hostile environment amplified by Google Maps' assumption of creating a total map.[75] Through rendering online racism like the White House incident a glitch, Google is able to abdicate individual responsibility for pervasive racism taking place online.[76] Racism online is either a temporary error in the system or the result of "bad actors" that are outside the system.[77] In either case, the assumption is that error can be contained.

Among the backlog of edits Map Maker was facing at its point of shutdown in 2015, we can assume that some of the edits were to report similar complaints of racist or misogynist information. The system overload made addressing these complaints untenable. As a result, these complaints are never directly addressed, at least through the line of recourse Google Maps sets up. Sara Ahmed argues that when institutions close the door on a complaints office (literally and metaphorically), the complaints do not magically disappear but are heaved onto the one making the complaint.[78] They must endure the persistence of that which needs to be addressed. So, with the shut-down of Map Maker we are left to wonder about the residue of error that continues to stain the map. What recourse is there for this? Where has the labour of content moderation moved and how are new infrastructures of participation like the Local Guides Platform of review moderated?

Terminating Google's Map Maker program did not address the issue of online racism and harassment but merely terminated one space where it took place. Map Maker's unofficial crowdsourced-mapping replacement, Google's Local Guides Platform,[79] maintains the same issues with spatial classification.

Indeed, harassment persists in the Local Guides infrastructure based on reviewing restaurants or rating customer service.[80] Local Guides becomes a platform to further codify space through metrics of investment; these are coded in language of authenticity, cleanliness, and friendliness; what then becomes located through the Local Guides program is the continued racist geographies.[81] By shutting down old crowdsourcing tools and replacing them with new ones—based on the same rationale to weed out the so-called bad apples—Google fails to acknowledge space as political, social, material formulations. Google's insistence on a world being technologically reproducible does not just assume a singular or linear representation of the world but also a failure to contend with the real harms that a narrow and ultimately hegemonic view of the world perpetuates. Sometimes covertly and sometimes overtly. Even if oriented as just a glitch, unjust geographies cannot simply be debugged by turning the machine off and on again.

The idea of the shutdown, as an action and as a failed strategy, demonstrates that Google Maps' errors are often not simple glitches. While Google Maps maintains a fantasy of total world knowledge and seamless mobility the glitch sheds light on the Google index of who belongs and who does not in the imaginary of its technological containment.[82] As an orientation of error, the shutdown distracts from the unjust geographies and default discrimination the map invests in and reinforces. Instead, the shutdown was an effort to recentralise the map to ultimately "serve Google's needs" and maintain Google's command of location awareness.

### Absence

While mapping reinforces processes of dispossession such as in the mapping of Buffalo, New York—thus absenting people and places from the map—absence from the map becomes an invitation to be mapped, as evidenced through the Map Maker tool. Therefore, Google Maps' constructed binary of absence and presence orients "empty spaces" and "missing data" as sites of potential growth and expansion. After Map Maker shut down, Google Maps developed new techniques of expanding spatial data collection, targeting the Global South.[83] Two recent Google Maps initiatives—*Tá no Mapa* (translated to English as "It's on the Map") and Google Maps' global Plus Codes project illustrate how Google Maps administers absence and presence on its map. *Tá no Mapa* implements a Street View-on-foot mapping of favelas in Rio de Janeiro, Brazil, while Plus Codes assigns "digitally sourced addresses" to the over one billion people in the world, who according to the terms of Google Maps, "don't have an address."[84] Google Maps' imposition of emptiness on that which it hasn't mapped imposes a colonial technique of claiming space in the name of "belonging to the map" through extracting data.[85] These two initiatives show how error for Google Maps is the double bind of both not being on the map and not being mappable. Specifically, space and place must be locatable on the map, because *accurate* spatial information is that which can be locatable. Google's work of inclusion is tied to being useful on the map whereby Google sets the terms and conditions of inclusion.[86] *Tá no Mapa* and Plus Codes reveal who Google considers as absent from the map and the techniques of inclusion Google Maps imposes.

In 2014, Google Maps turned its attention to Rio de Janeiro, Brazil. Rio was preparing to host the FIFA World Cup of soccer that summer, as well as preparing for the upcoming Summer Olympic games in 2016.[87] Google was among the other big companies ready to invest in Rio's "infrastructural development" with an interest in Rio tied to the surge in global tourism about to descend on the city. Mega-event hosting is often a boon to planning and development, while at the same time, such a surge in building exacerbates inequity through processes of forced removal.[88] In the case of Rio's development, these include the speedy construction of a rapid transit system through the city[89] that disproportionately affected people living in Rio's favelas.

While development projects treat favelas as removable, "informal," and erasable settlements, favelas are densely populated and intensely infrastructural areas of Rio, home to 25% of Rio's population.[90] Many people living in favelas have faced historical racist and classist structures of dislocation and dispossession. While the area is stereotyped as dangerous, a 2013 study shows that 85% of favela-residents are happy with where they live.[91] Such narratives of danger and poverty make them susceptible to redevelopment, especially with the impending mega-events where favelas like Villa Autódromo were destroyed to free up land for redevelopment—proximity to the water and to Olympic Park.[92]

Within the displacement of mega-events and long-term planning for short-term tourist-based infrastructures, Google was navigating its own limitations. There had been a growing number of reports of tourists using Google Maps in Rio and finding themselves, *by accident*, in the city's favelas. The trouble was most favelas did "not exist" according to Google Maps—or at least they were not mapped by Google Maps. Instead, they were marked as blank "grey" spots on Google Maps suggesting nothing was there. Their absence on the map was attributed to the favelas not having wide enough roads for the Google Street View car to drive down. With the eminent influx of international tourism and attention in Rio, thanks to the World Cup and the Summer Olympics, Google also launched other interactive "cultural experiences" including 360-degree "motorbike" tours through the city's alleys. "With any luck, Google's project and the 2016 Olympics will leave a lasting and positive legacy on the sprawling slums."[93] While favelas were projected as inhospitable to Google Maps' global mapping ambitions, Google Maps ascribes new strategies for mapping the space.

Thus began the *Tá No Mapa* or On the Map project. Here Google partnered with Afroreggae, an art collective from Rio to add details to what Barroso might call static areas.[94] *Tá No Mapa* mapping strategy is a variation on the Street View mapping processes. Rather than attached to the official

Street View Car, Google's 360-degree cameras are attached to a special backpack to be worn by local Google Maps' "field agents." These field agents are tasked with walking through Rio's favelas wearing these cameras on their backs and documenting images in 360-degree view. According to Google's project description, Google selected locals based on their knowledge of the city, specifically details like the short cuts and the alleyways otherwise inaccessible to street-view vehicles.[95] Google wanted to make their sensed and embodied knowledge legible to Google Maps users. Google's process of local buy-in intentionally obscures Google's power. With the appearance as a broad-based community initiative, not only does Google Maps profit from the expertise of the "local informant" it also presents itself as coming from the inside, which is at least the perspective of the Western press.[96] In the case of the racialised other, these processes are seen to be of great benefit to the community, connecting them to state infrastructures and the world outside the favela.[97]

To add to the documentation of the *Tá No Mapa* project, Google produced a series of YouTube videos about the project. This video series, presented on the Google Maps YouTube channel, documents the narratives and routes of the favela residents who are part of the data collection process.[98] These videos follow people walking with Google cameras strapped to their backs but moving through their neighbourhood as they talk with their neighbours and wave at shop owners. One video follows Afroreggae volunteer José Junior walking through a favela with the camera on his back. As he walks through the streets, he explains to the audience that what he is "mapping" with the camera would otherwise not be on the map. He approaches the video as if he is giving a private tour of his neighbourhood, emitting pride in where he lives—the Parada De Lucas favela, home to over 20,000 people. The video introduces this community as a place that "up until now didn't exist" since it was not on Google Maps. Other videos in this series include *Luis' Story* about a ballet dancer living in the favela.[99] There's also *Ricardo's Story* about a surf instructor from Rocinha,[100] and *Paloma's Story* about a student studying computer science with the ambition of working in the tech sector.[101] These personal narrative videos augment the mapping process. The overarching theme of the videos is community-building rather than data, yet the language employed by Google is that of record-keeping and indexing. In other words, the pretext is that this is for the people to benefit from Google's platformisation of place that has, up until now, represented Rio's favelas as an "uncharted and mysterious place on the map."[102] Google constructs the terms of legibility in who and what is made legible and worthy of being mapped.

Through *Tá No Mapa*, Google Maps renders favelas a product of the map where Google Maps decides the terms of inclusion. Google's commerical terms are reinforced through Afroreggae's proclamation that this mapping project is "more than inclusion but to show businesses" and "to enable these people to be part of society as a whole."[103] The selection of businesses is included in

Google's mapping of areas of interest. Critical Geography scholars Andrés Luque-Ayala and Flávia Neves Maia[104] identify that the Google Maps' category of "areas of interest" is a deeply political label and the classification metrics are opaque. As with the Fruit Belt neighbourhood, Google's Areas of Interest are financially driven labelling of space—a means of highlighting commercial areas on the map. Areas of Interest not only decide what is "of interest" on the map but also reference a North American planning logic of commercial streets buttressed by residential areas, a formation that is not universal. What Google fails to account for is the basis for this inclusion—in other words, how they decide who is to be included and on what terms and the processes that determine what makes information valuable to Google. As such, Google Maps effectively depoliticises inclusion when it fails to acknowledge their own term through how people and places are brought into their global map. Instead, *Tá No Mapa* is a form of racial capitalism that presents *diversity* "as a business good."[105]

Delineating what counts as an area of interest isn't necessarily decided on by local data inputters but is, instead, filtered through processes of image analysis and interpretation. Google's own in-house machine learning systems and manual inputters analyse and interpret the images. For Google this is a form of digital visibility, the map breaks through the wall between the favela and the city. However, for Luque-Ayala and Neves Maia *Tá No Mapa* homogenises the assemblages of the favela and reifies "the very walls that the digital map is claiming to break."[106] Instead, *Tá No Mapa* is a form of racial capitalism that presents *diversity* "as a business good." Google controls the terms through which favelas are seen, flattened to look like the same business areas and transactional spaces of capital.

Further, the connection with the "outside world," and feeling as part of a community, appears uni-directional, for the benefit of (and contingent on) outsiders coming into the favela. The map promises an opening of the favelas to more explorers, more travellers, who now interpret the favelas as legible to them and a place to consume. The other side of this invitation is to bring an audience to the map to see the favelas—or to "see them in 360-degrees for the first time" as the Google video suggests. In this sense Google penetrates the opacity of the favela, an opacity that in many ways can also serve as protection. Through Google's impositions of absence, error is anchored to colonialist data regimes emboldened by the colonialist imaginary of boundless access to all space in the name of making space "knowable."

Google operates similar mechanisms of inclusion in their global Plus Codes program. Plus Codes is a Google Maps project that assigns addresses to people who do not have addresses in the normative sense of fixed location. Google defines Plus Codes as "digitally sourced addresses" based on Google's proprietary grid system. Plus Codes are only visible and readable through Google Maps and according to their system.[107] These addresses are based on the geographic coordinate system of latitude and longitude, coordinates reinterpreted through a proprietary alphanumeric system of the "digital globe." In other words, Plus Codes appear similar to decimal degrees of latitude and longitude used in many Graphic Information Systems (GIS) and GPS (Graphic Positioning System), but, they are a reinterpretation of this familiar system, translated to a private language of 20 alphanumeric characters. Through Plus Codes, Google Maps can ascribe a location to the size of a 3m × 3m zone or what they cite as "an exact address for a sidewalk vendor who may not even have a storefront."[108] The grids, or "tiles," are divided and labelled and then further divided and labelled, repeated to build a full code. Codes read like 5G5CW2GJ+JQ for the Google Johannesburg Office. The codes are a digital grafting of space, always in relation to Google Maps and do not translate to the world beyond their apparatus.

David Martin, Director of Program Management at Google Maps' introduced the Plus Codes program with the declaration "we're giving everyone, everywhere an address."[109] According to Google, Plus Codes have been assigned in "The Gambia, Kenya, India, South Africa and the U.S., with more partners on the way." There are reports from NPR of "Navajo Nation Homes Get Addresses from Google Mapping Project" having assigned more than 500 signs displaying Plus Codes by the Rural Utah Project in San Juan Country.[110] This is in the name of "voter registration." Like the *Tá No Mapa* project's "it's on the map," Plus Codes frames itself as "helping people get on the map" by directly mapping people.

According to Google, Plus Codes prevent "error" by avoiding the use of characters that could be confused like "1," "L," and "l." Plus Codes do not use vowels to prevent accidental word formations. Google assumes that its alphanumeric system is a universal coding system that can "be used by anyone no matter what language you speak." Of course, this is a specific Roman character system that is not used by many of the so-called billions of people "who don't have an address." Not to say that it won't be understood, but it is an extension of Google's English-dominant web presence.[111] The choice of characters alone exposes the reality that being put on the map does not advance the interests of the people being mapped but advances the interests of Google in a language not only understood by Google, but in a configuration of language owned by Google.

At the same time, Google Maps promotes Plus Codes as an everyday navigational aid and "a way to share any location—your home, a store, a meeting place—even if there are no street names or even streets. Friends, family, and service providers will find your home without relying on landmarks or asking for directions along the way."[112] It is also a travel aid for "When you travel to unfamiliar places" as a means to "save you time by guiding you directly to your exact destination." Plus Codes' promises reflect the general promises of personal optimisation discussed in Chapter 4. They are a means to move efficiently through space, to know where to go. The Plus Codes promise of precision supersedes the other spatial organisational practices and embodied experiences of place. In other words, Google Maps is literally fixing people to their propriety mapping infrastructure (Figure 5.2).

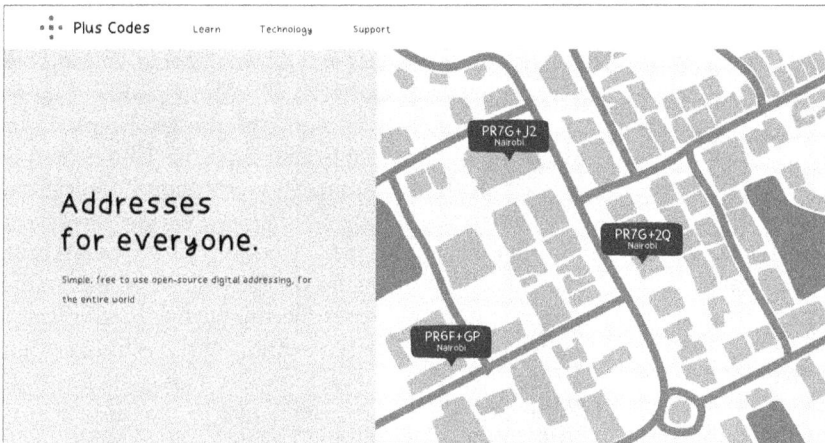

*Figure 5.2* Illustration of the Plus Codes Home page with the title: "Addresses for everyone. Simple, free to use open-source digital addressing, for the entire world." The image is the Google Maps satellite view of Nairobi with Plus Codes pinned to the map.

But Plus Codes are more than simply a navigational aid. Like *Tá No Mapa*, Plus Codes are used as what Luque-Ayala and Neves Maia identify as a "mechanism for socio-economic inclusion."[113] For Google, creating location codes is a means to "access things like banking and emergency services, receive personal mail and deliveries, and help people find and patronise their businesses,"[114] thus marking the terms of inclusion as participation in financial (banking) and logistical (delivery) networks of capital. Google Maps further entrenches itself as core infrastructure of capital flows—even if one's location is not legible through Western terms of ordering space such as a street number, a proximate intersection, or a post-code. The Plus Codes enforce a technique of mapping based on Western locational ordering according to property but are extrapolated to be a universal index, applicable everywhere globally, and moreover, the process of mapping is a process owned by Google Maps.

While Google Maps promotes Plus Codes as being open source and free to use, we know that they are part of a proprietary, infrastructural "free open-source algorithm." But what are the hidden costs of open source? Google Maps is entrenching itself in capital flows not just by giving people addresses, but by giving people addresses that can only be used in Google Maps' system. But this circulation is far from free as it ensnares individuals in a system, enforcing regimes of data inclusion and data legibility,[115] reinforced through what Nawal El Saadawi and others have argued are imperialist development infrastructures like Western "aid" or NGO "development" initiatives.[116] Plus Codes uses Google Maps' infrastructures here with the promise that Plus Codes will help people receive packages, have access to bank accounts, and even vote.[117]

Embedded in the assimilation process is the thinking that people who live without stable or settled addresses in places that are geographically governed by addresses (in the sense of discrete street names, sequential numbering systems, and postal codings), like North American cities for example, do not already have alternative solutions or networks to provide them with places of access. Instead, inclusion is an act of tethering to the map. Plus Codes reveals the predatory nature of Google's location awareness—where people become the currency of precision. While these types of data extraction processes have been labelled data colonialism or digital colonialism,[118] Roopika Risam reminds us that data cultures are already deeply colonial and the imposition of imperialist data to organise and manage is not unique to the digital age.[119] Similarly, Leanne Betasamosake Simpson argues that "extraction and assimilation go together. Colonialism and capitalism are based on extracting and assimilating."[120] Plus Codes is both extractive and assimilative in the sense that it allows Google to extract value from the processes of mapping and the people it literally codes into (or assimilates into)—Google Maps. It thereby reinforces the role of data as a colonialist tool that serves the logistics of hegemony. It helps power administer, manage, and demarcate at the scale of the individual.

The inclusion narratives of Plus Codes reflect what Hoffmann argues is a "discursive violence of inclusion" which Hoffmann identifies as narratives that run amok in discourses about data, especially in relation to ideas of inclusion.[121] As Hoffmann writes, this is the kind of violence "that operates by diffusing resistance, deepening dependency on oppressive structural conditions, and preserving the potential for other forms of violence."[122] In other words, by promising the delivery of access to such essential processes as emergency medicine and voting, Google not only situates itself as a tool of survival and citizenship, it becomes impossible to challenge those goals. Operating at the scale of individual access, it leaves out the reality that Plus Codes are part of the systems that help power administer, manage, and demarcate belonging in the first place.

Moreover, the process of being mapped presumes visibility, and more specifically visibility as a form of inclusion, as something that is always desired. As Rianka Singh and Sarah Banet-Weiser argue, "the empowerment of the platform, as it rests on becoming and being visible and noticeable, also means that bodies—often abject or marginalised bodies—are put on display and made vulnerable to a range of surveillance mechanisms."[123] The Plus Codes project consolidates visibility on and to the map in the service of local governance, NGOs, and administrators. It presumes this form of inclusion is desireable. But, being visible to (and according to the terms of) technical systems, to governments, or to administration is not always safe.[124] Being given an address, or literally put on the map via an alphanumeric code, demonstrates the competing goals of being seen and being found. As Singh and Banet-Weiser contend, "visibility can be a trap."[125]

Google programs like *Next Billion Users* and The Google Station—its internet infrastructure building project that targets cities in India, Mexico,

Indonesia, Philippines, Thailand, and Nigeria—demonstrate a pervading focus on expanding the reach of its core products, like Maps, and "mine the market potential" of the Global South.[126] These colonial projects not only target the idea of Google users but also of Google workers available to add and moderate spatial information as well as expand Google Maps' coverage of experience. This form of spatial possession (and its corollary of dispossession) ignores the relationality of place and instead enacts Kim Reynolds's concept of "extraction as white supremacy."[127] Here I am thinking with Leanne Betasamosake Simpson's framing of extraction as a "cornerstone of capitalism, colonialism, and settler colonialism. It's stealing. It's taking something, whether it's a process, an object, a gift, or a person, out of the relationships that give it meaning, and placing it in a nonrelational text for the purpose of accumulation."[128]

*Tá No Mapa* and Plus Codes demonstrate how Google Maps forecloses mapping "accuracy" as the "colonization of the communication space in the Global South shapes the essence of digital coloniality is a form of imperialist capitalism."[129] Here the investment in space is not in urban development or defining routes, but specifically an investment in people as locatable objects and people as located objects. This encoding of places and people is framed as a social service, reorienting attention towards the possibilities rather than the problems. Like digital redlining's amplification in the racist investment of structures of city planning and ownership, digital colonialism capitalises on already present colonial infrastructures as part of the ongoing processes of "economic and political domination."[130]

Google Maps' orientations of error—a temporary fault line to be corrected or a point of absence that can be filled—demonstrate how Google's location awareness is a form of location indexing based on systems of racial capitalism. This is the muddy mess of reorientation that is lost in the depoliticised despair of losing a sense of direction to digital mapping platforms. Therefore, when we think of Google Maps' orientation of error and the magnetic response to simply write mistakes off as glitches, we must follow Benjamin's line of inquiry and ask, *how is technology already raced?*[131] When doing so, we can reread the headlines at the start of this chapter and understand that conditions of Google's spatial ontology of location awareness are more than just a bog masquerading as a viable route to the airport. In mapping Google Maps' territory of error, we are oriented to Google's conception of the simple glitch, "going nuclear" and addressing absence as a process of dispossession rather than inclusion. While Google's orientations of error are directed by commercial interests, it is how these interests are constitutive of enduring forms of racial capitalism and colonialist extraction that remain the heart of Google Maps' enclosures. When Google Maps invests in the data regimes of redlining and imperialism, the errors it makes are more than about navigating place but are also about the types of overdetermined relations the technology carries with it and amplifies.

Despite Google Maps' discursive enclosures of place, using Google Maps does not make one determined by the map. We reroute ourselves, go our own way, and follow our own embodied sense of direction. Google's conditions of accuracy are not a matter of individual spatial rhythms, and Google's errors cannot be outpaced by leaving a few minutes early to make sure the directions are correct. Google's conditions of accuracy are the conditions and structures of unjust geographies built into mapping infrastructure in a project of creating a totalising map. The "wrong way" is more than just a detour we navigate, it is the swamp of the legacies of data systems folded into everyday experiences of place.

Collapsing location awareness into the despair that *we don't have a sense of direction anymore* or that we are *too reliant on Google Maps* is itself disorienting an understanding of what location awareness is even in the age of Google Maps. This condemnation undergirds Kastrenakes' annoyance as well as much of the reporting on navigational mishaps. But this line of thinking distracts from the reality that we all carry with us a sense of direction. A sense of direction is an ongoing relationship with place. It is a situated practice informed by location, identity, and power, and its signifiers such as gender, race, class, sexuality, ability, and borders. This is what Sara Ahmed calls the "starting point for orientation" defined as "the point from which the world unfolds: the "here" of the body and the "where" of its dwelling.[132] As discussed in Chapter 3, Google Maps configures orientation of use through processes of white prototyping.[133] These conditions of prototyping translate to Google's orientation of error—and form the basis through which they establish what directions are right and what ones are wrong. But as always, there is more at stake than simply going the right way.

**Notes**

1 Sweeney, "Google Maps Sends Hikers"; Mountaineering Scotland, "Google Maps Risk"; John Muir Trust, "Online Routes."
2 Kennedy, "'Muddy Mess' in Colorado"; Terez and Mielke, "Google Maps Shortcuts."
3 Suarez, "Google Maps Mistakes."
4 Hough, "US Woman Sues Google."
5 Cuthbertson, "Google Maps Error."
6 Sutter, "Google Maps Border." CNN's reporting reveals a not-so-coded imperialism when reporting on the storied Costa Rican and Nicaraguan border tension and makes light of the siutation.
7 Kastrenaskese, "Do Not Blame Google Maps."
8 Lloyd, "Information Literacy Landscapes," 570–583; Lloyd, "Framing Information Literacy," 245–258; Lloyd, *Information Literacy Landscapes.* While traditional media and information literacy might contend that such mistakes arise from a deficit of skill or savviness when reading information, with the correction being to teach the skills of media savviness to better interpret information.
9 Reid, "15 Years of Mapping."
10 Google, "Additional Terms of Service."
11 Google Maps, "Why We Map."
12 Vertesi, "Mind the Gap," 7.
13 McKittrick, *Demonic Grounds.*

14 Benjamin, *Race After Technology*, 28–29.
15 "Then Google Maps was like, 'turn right on Malcolm ten Boulevard' and I knew there were no Black engineers working there." Allison Bland (@alliebland) quoted November 19, 2013 in Benjamin, *Race After Technology*, 78.
16 Benjamin, *Race After Technology*, 78.
17 Ali, "Malcolm X in Brooklyn."
18 Benjamin, *Race After Technology*, 80.
19 Russel, *Glitch Feminism,* 7.
20 Russel, *Glitch Feminism.*
21 Berlant, "The Commons," 393.
22 Berlant, "The Commons," 393.
23 Bethel, "The Fruit Belt Demonstrates"; Buckley, "What's in a Name?"; Dewey, "Fruit Belt Fights."
24 Dewey, "Fruit Belt Fights."
25 Fruit Belt Community Land Trust.
26 Fruit Belt Community Land Trust.
27 Fruit Belt Community Land Trust.
28 Fruit Belt Community Land Trust.
29 Bethel, "The Fruit Belt Demonstrates."
30 Doig, "Get more power?"; Epstein, "Medical Campus Development Boom"; Dewey, "Fruit Belt Fights."
31 Buffalo Niagara Medical Campus, "Entrepreneurship."
32 Fruit Belt Community Land Trust.
33 Dewey, "Fruit Belt Fights."
34 Noone and Jacob, "Bad Boundaries," 76–92.
35 Dewey, "Fruit Belt Fights."
36 Dewey, "Fruit Belt Fights."
37 Robinson, *Black Marxism.*
38 McKittrick, *Demonic Grounds,* x.
39 Fruit Belt Community Land Trust.
40 McKittrick, *Demonic Grounds*, xi.
41 Bhattacharyya, *Rethinking Racial Capitalism*, 102.
42 Gilliard, "Pedagogy and the Logic of Platforms"; Noble, *Algorithms of Oppression*; Safransky, "Geographies of Algorithmic Violence," 200–218.
43 Light, "Discriminating Appraisals," 485–522.
44 Shapiro, "Street-level: Google Street View's abstraction by datafication."
45 Safransky, "Geographies of Algorithmic Violence."
46 Lisa Nakamura quoted in Barrett, "Google Maps Is Racist."
47 Noble, *Algorithms of Oppression*, 1.
48 According to Google Maps' Mark Li and Zhou Bailiang, "… you'll notice areas shaded in orange representing 'areas of interest'—places where there's a lot of activities and things to do. To find an 'area of interest' just open Google Maps and look around you. When you've found an orange-shaded area, zoom in to see more details about each venue and tap one for more info. Whether you're looking for a hotel in a hot spot or just trying to determine which way to go after exiting the subway in a new place, 'areas of interest' will help you find what you're looking for with just a couple swipes and a zoom. We determine 'areas of interest' with an algorithmic process that allows us to highlight the areas with the highest concentration of restaurants, bars, and shops. In high-density areas like NYC, we use a human touch to make sure we're showing the most active areas" (Li and Bailang, "Discover the Action.").
49 Hawkins, "The Deep Dive."

50 O'Beirne, "Google & Apple Maps"; O'Beirne, "What Happened to Google Maps"; O'Beirne, "Google Maps's Moat."
51 Sharma, "Introducing Power-Chronography," 67.
52 McKittrick, *Demonic Grounds*, xviii.
53 Zuboff, *Age of Surveillance Capitalism*.
54 Here I am invoking Legacy Russell's *Glitch Feminism*.
55 hooks, "Choosing the Margin," 15–23.
56 Nicas, "Google Maps Renames Neighborhoods."
57 Nicas, "Google Maps Renames Neighborhoods."
58 Plantin, "Google Maps as Cartographic Infrastructure"; Loukissas, *All Data are Local*.
59 Benjamin, *Race After Technology*, 78.
60 Elwood, "Digital Geographies," 216.
61 Gillard, "Pedagogy and the Logic of Platforms," 64.
62 Brock, *Distributed Blackness*, 134.
63 I discuss Ground Truth in Chapter 2.
64 Hern, "Google Maps Hides an Image."
65 Google, "Google Maps/Google Earth Additional Terms of Service."
66 Pritchett, "Fake and Fraudulent Content."
67 Noble, *Algorithms of Oppression*.
68 Barrett, "Google Maps Is Racist."
69 Hern, "Google Maps Hides an Image."
70 Mosendz, "Google Limits Access"; BBC News, "Google Suspends Map Maker."
71 Kuo and Marwick, "Critical Disinformation Studies"; Nakamura, "Glitch Racism."
72 Pichai, "HBD Maps!"
73 Gilmore, "Fatal Couplings," 15–24; McKittrick, *Demonic Grounds*.
74 Charlton McIlwain and Lisa Nakamura in Barrett, "Google Maps Is Racist."
75 Hoffmann and Jonas, "Recasting Justice."
76 Nakamura, "Glitch Racism."
77 Kuo and Marwick, "Critical Disinformation Studies."
78 Ahmed, *Complaint!*
79 Local guides are discussed in more detail in Chapter 3.
80 Bhandari and Noone, "Support Local."
81 Zukin, Lindeman, and Hurson, "The omnivore's neighborhood?"
82 Plantin, "Google Maps as Cartographic Infrastructure," 500.
83 Magalhães and Couldry, "Giving by Taking Away," 343–362; Oyedemi, "Digital Coloniality," 329–343.
84 Google, "Plus Codes."
85 Tuck and McKenzie, *Place in Research*.
86 Gangadharan, "Downside of Digital Inclusion," 597–615.
87 Chester, "How Google Put Rio's Favelas."
88 Kassens-Noor et al., "Mega-Event Legacy Framework"; Leopkey and Parent, "Governance of Olympic Legacy," 1–14; Smith, "Leveraging Sport Mega-Events," 15–30.
89 Kassens-Noor et al., "Olympic Transport Legacies," 13–24.
90 Matchar, "Mapping Rio's Favelas."
91 Carta Capital, "Brazil's 5th Largest State."
92 Talbot, "Vila Autódromo."
93 Chester, "How Google Put Rio's Favelas."
94 Google, "Beyond the Map."
95 Matchar, "Mapping Rio's Favelas."
96 Chester, "How Google Put Rio's Favelas."

97 Milan and Treré, "Big Data," 319–335.
98 Google, "Beyond the Map."
99 Google, "Luis' Story."
100 Google, "Ricardo's Story."
101 Google, "Paloma's Story."
102 Luque-Ayala and Neves Maia, "Digital Territories," 458, quoting Google, "Beyond the Map with Google."
103 Jose Junior in Google, "Mapping the Favelas."
104 Luque-Ayala and Neves, "Digital Territories," 449–467.
105 Bhattacharyya, *Rethinking Racial Capitalism*.
106 Luque-Ayala and Neves Maia, "Digital Territories," 457.
107 Bishop, "A Simple and Accurate Address"; Google, "Addresses for Everyone."
108 Martin, "Everyone, Everywhere an Address."
109 Martin, "Everyone, Everywhere an Address."
110 Sevigny, "Navajo Nation Homes."
111 Mpofu and Salawu, "African Language Use," 76–84; Kadenge and Nkomo D., "English Language in Zimbabwe," 248–263.
112 Google, "Learn about Plus Codes."
113 Luque-Ayala and Neves Maia, "Digital Territories," 450.
114 Pichai, "HBD Maps!"
115 Mervyn, Simon, and Allen, "Digital Inclusion," 1086–1104; Milan and Treré, "Big Data," 319–335; Gangadharan, "Downside of Digital Inclusion," 597–615.
116 El Saadawi, "Dissidence and Creativity," 1–17.
117 Google, "Addresses for Everyone."
118 Couldry and Mejias, *The Costs of Connection*.
119 Roopika, *New Digital Worlds*.
120 Tuck and McKenzie, *Place in Research*, 67.
121 Hoffmann, "Terms of Inclusion," 3539–3556.
122 Hoffmann, "Terms of Inclusion," 3540.
123 Singh and Banet-Weiser, "Sky High: Platforms and the Feminist Politics of Visibility," 165.
124 Spade, *Normal Life: Administrative Violence*.
125 Singh and Banet-Weiser, "Sky High: Platforms and the Feminist Politics of Visibility," 165.
126 Oyedemi, "Digital Coloniality,"
127 Kim Reynolds, "Extraction as White Supremacy."
128 Simpson, *As We Have Always Done*, 213.
129 Oyedemi, "Digital Coloniality," 333.
130 Oyedemi, "Digital Coloniality," 337.
131 Benjamin, *Race After Technology*, 79.
132 Ahmed, *Queer Phenomenology*, 8.
133 Browne, *Dark Matters*.

# 6    Epilogue

## Reorienting Location Awareness

Having come this far into the woods of location awareness, a tempting question might be: Is there a way out? Is there a solution to make Google Maps better? Or is there a better tool to use instead of Google Maps?

In the Grimm Fairy Tale, Hansel and Gretel, Hansel lays out bread crumbs as he and Gretel wander further into the darkness of the forest. Bread crumbs mark a provisional pathway in unfamiler terrain. They are an ad hoc mapping of a route, a trail of location awareness, to track a path to *get out* the same way they got in. The questions—*Is there a solution to make Google Maps better? Or, is there a better tool to use instead of Google Maps?*—operate similarly to Hansel's bread crumbs. They continue to charge forward without attending to the context of these locative tools. What benefit does a "better map" have when the direction of universal mapping is the same? As the brothers Grimm have it, the bread crumbs are, indeed, eaten and the *way out* becomes more difficult to find.

Throughout the book I argue that Google Maps' project of building location awareness is a project nested in other imaginaries of public good, spatial entitlement, contained cities, and presence as always good. But more than simply selling the promise of a total map, the single map is offered as a public resource, an aid to self-sufficient exploration, a means to render legible the chaos of space, and a platform to be found on. In selling a global mapping project through these registers of spatial relation, Google Maps effectively delineates the contours and passages of publicness, access, readability, and error. But more than that, the total mapping of the world swallows whole the markers of space that make it uninhabitable for many people in the first place. The stratified and dispossessory organisations of space are the territory that Google's location awareness both absorbs and sustains. Instead of the question "is there a solution?" I revisit John Brian Harley's question: *can there be a cartographic ethics?*[1] Or, applied to conditions of Google Maps: is there an ethical way to map the *entire* world in the name of location awareness?

The desire may be to fix Google Maps with a *better* design or *additional* data or *more* means for public input. These fixes come packaged as "data ethics" or "inclusive design." But as Anna Lauren Hoffmann argues, such banners distract from the violence of these systems of classification and sorting.[2]

DOI: 10.4324/9781003251569-6

Hoffmann writes how hackneyed narratives of data ethics and inclusive design "represent what we might call the discursive excess of inclusion; inclusion discourses do not simply normalise, but dupe us into celebrating the very power structures that generate asymmetrical vulnerabilities to violence in the first place."[3] Imagining a way to "add ethics" to a project of world mapping is itself disorienting when the foundations of such a project are based on impulses to know and to claim the world as one's own.

Google's attendant mapping projects like Project Air View to Plus Codes, where Google leverages Maps as a social good, reify the asymmetries of space, circumscribed to the maps' surface. Where is the room, or the space, for ethics in such acts of containment and branding? Administering space via a global mapping mission enforces a knowing of space and a besting of space. It both contains and flattens its depths, complexities, pleasures, and hostilities.[4] Google is not the first and only attempt to manage space in this way. Administering space via the map is tied up in its own histories of command and control, delineated through colonial boundaries, redlining, GIS precision. These have lived consequences beyond the map, that translate to how space is valued and claimed. These precedents ground Google Maps, a grounding made to seem banal (while no less harmful) through a project of location awareness.

Then there is the desire to look for a better tool—one that works better, is more fair, less biased, *more democratic*. Indeed, part of what makes the universal map seem so appealing is the idea that anyone can map or that anyone can have a sense of direction that can be supported or augmented through the location awareness affordances of the universal map (along with real-time feedback). Alternatives like OpenStreet Map based on crowd-sourced information, or Apple Maps, which promises no location tracking, present as the ethical alternative. However, as Sarah Elwood and Agnieszka Leszczynski as well as Renée Sieber and Mordecai Haklay have argued, volunteered or crowdsourced geographic information does not necessarily make mapping more open; but instead, volunteer mappers become co-authors in these hegemonic spatial ontologies.[5] These mapping projects must also address the pressing questions of who is doing the work of mapping on these platforms, who do these maps serve, and who wants to be seen in the first place? Alternative platforms like Apple Maps or OpenStreet Map make claims to privacy, yet operate through the same vision of a total map.

The framework of "opting out" of Google Maps—to choose not to use Google Maps—leaves out the complexity and reach of these mapping systems that extend beyond use. To be part of the map is to be made consumable by the map; and, to not be *on the map* is to be erased from the map and the terms by which space is made investable. The conditions of inclusion play out in places like the Fruit Belt neighbourhood in Buffalo, New York. The Fruit Belt neighbourhood didn't choose to have their name replaced by the same development they have been fighting for decades, nor did they accept their absence from the Map. By contrast, when Stephen Petrow from Chapter 1 sets out to travel across America without using Google Maps, he fails to recognise that he is already the model and operation of the map's location awareness. And while it is no longer a surprise that these technologies and tools are racist, classist, sexist, and make space dangerous for queer and trans people, what Google's mapping projects make clear is how the prototyping of hegemonic norms is and has always been a spatial project.[6] Hegemony is not simply routed throughout Google Maps; it is Google Maps' territory.

The conditions of Google Maps' location awareness extend beyond using the map and indeed precede the map and make the imaginary of a productive universal vision possible, at the expense of all that does not fit. So while Ground Truth, Air View, Explore Pages, Immersive Views, and Plus Codes may operate through seemingly precise and practised location awareness, they rely on an inattention to how they organise and orient space according to where is risky and where is claimable, and what is smoothed out and what is included. But these are not the only terms of location awarenss. Beyond the map, location awareness is particularistic and corpuscular—it is practised survival, it is impromptu encounter, and it is fleeting reverie. Refusing to follow the path of "we don't have a sense of direction anymore," unearths a myriad of other possible ways that are activated and reactivated with and without Google Maps every day (Figure 6.1).

*Figure 6.1* Textures of location awareness on the streets of London. Photograph by Author, 2018.

Returning to the metaphor of the breadcrumbs and the promise of a way out, this book traces a number of trails already set in response to Google's mapping of space. These are pathways to a way out created by people who are, in big and small ways, working against Google's mapping project, or reworking Google's tools to tell a different story about location awareness. These are the trails that lead to a challenging of the map. Challenge comes in the form of confronting Google Maps' terms of inclusion, such as the Fruit Belt fighting to be on the map as a means to assert their presence amidst Buffalo's redevelopment initiatives. It is Mapping Access's intervention into Google's compliance-based framing of access, adding and annotating spatial information about experiences of space from a variety of perspectives. It is also playing with the information Google Maps provides, such as Robin Maynard's charting of an alternate spatial intelligence using Google Maps'

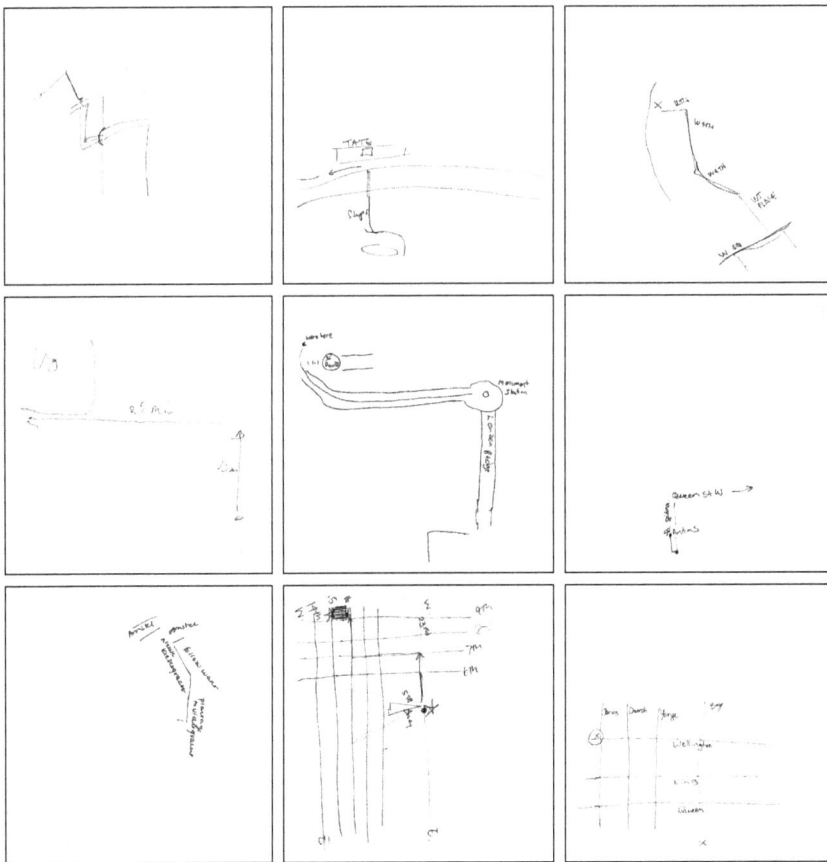

*Figure 6.2* Nine examples of direction drawings collected by the author from Amsterdam, London, New York City, and Toronto in 2017–2018, inspired by Stanely Brouwn.

routing function, to make visible the companies at the helm of environmental destruction and extraction that otherwise blend into cityscapes. Or there is Tamiko Thiel's artwork that uses the digital map's locational coordinates to display big tech's carbon output over their Silicon Valley headquarters. Refusing Google's location awareness is also refusing that which happens outside the map, such as grassroots organising in Oakland California against the capitalist systems that make Oakland a target. These are the ways that people refuse Google Maps' organisation and orientations of the world.

To lead us out, we return to Stanley Brouwn, my starting point. Brouwn continued to quietly perform *this way brouwn* until his death in 2017. Never one for interviews or celebration, Brouwn's work recedes into the background of the everyday while nonetheless providing a through line from the early 1960s to today, the Age of Google Maps. In that sense, Brouwn also performed his work of "discovering streets" alongside Google's location awareness—continuing the act of collecting drawings and asking for directions. The quiet performance of being lost dwells in the incomplete knowledge of the world. And while it could be easy to slip back into the fantasy of a life without Google Maps in Brouwn's ever unfolding circuity, a celebration of ditching the phone and asking for directions, Brouwn's work remains a provocation into the futility of measuring, claiming, or defining space. Instead, location awareness unfolds along multiple axes, *this way*:

"Walk during a few moments very consciously in a certain direction; simultaneously an infinite number of living creatures in the universe are moving in an infinite number of directions"[7] (Figure 6.2).

## Notes

1 Harley, "Can There Be a Cartographic Ethics?" 9–16.
2 Hoffmann, "Terms of Inclusion: Data, Discourse, Violence," 3539–3556.
3 Hoffmann, "Terms of Inclusion: Data, Discourse, Violence," 3550.
4 For more on mapping and surfaces, see Massey, *For Space*.
5 For example see: Sieber and Haklay, "The Epistemology(s) of Volunteered Geographic Information: A Critique"; Elwood and Agnieszka Leszczynski, "New Spatial Media, New Knowledge Politics," 554–559.
6 For example see: McKittrick, *Demonic Grounds*; Simpson, *As We Have Always Done*; Tuck and McKenzie, *Place in Research*.
7 Russeth quoting Brouwn in *Art & Project*, 1969; Russeth, "Stanley Brouwn."

# Bibliography

Aclima. "Built for Good." "About Us" page. Accessed October 11, 2023. https://www.aclima.io/about.

Ahmed, Sara. "Institutional as Usual: Diversity Work as Data Collection." Video of lecture given at Barnard College, filmed October 16, 2017. https://feministkilljoys.com/2017/10/24/institutional-as-usual/.

Ahmed, Sara. *Complaint!* Durham, NC: Duke University Press, 2021.

Ahmed, Sara. *What's the Use? On the Uses of Use*. Durham, NC: Duke University Press, 2018.

Ahmed, Sara. *On Being Included: Racism and Diversity in Institutional Life*. Durham, NC: Duke University Press, 2012.

Ahmed, Sara. *Queer Phenomenology: Orientations, Objects, Others*. Durham, NC: Duke University Press, 2006.

Ali, Zaheer. "Malcolm X in Brooklyn." *Black Perspectives*. February 20, 2017. https://www.aaihs.org/malcolm-x-in-brooklyn/

Anderson, Benedict. *Imagined Communities: Reflections on the Origin and Spread of Nationalism*. London, UK: Verso, 1983.

Ash, Johnathan, Rob Kitchin, and Agnieszka Leszczynski. "Digital Turn, Digital Geographies?" *Progress in Human Geography* 42, no. 1 (2018): 25–43.

Bagli, Charles V. "$2.4 Billion Deal for Chelsea Market Enlarges Google's New York Footprint." *New York Times*, February 7, 2018. https://www.nytimes.com/2018/02/07/nyregion/google-chelsea-market-new-york.html.

Barnett, Anna L., Mellissa Prunty, and Sara Rosenblum. "Development of the Handwriting Legibility Scale (HLS): A Preliminary Examination of Reliability and Validity." *Research in Developmental Disabilities* 72 (January 2018): 240–247.

Barns, Sarah. *Platform Urbanism: Negotiating Platform Ecosystems in Connected Cities*. London: Palgrave, 2020.

Barrett, Brian. "Google Maps Is Racist Because the Internet Is Racist." *Wired Magazine*, May 23, 2015. https://www.wired.com/2015/05/google-maps-racist/.

BBC News. "Google Suspends Map Maker Because of Vandalism." May 12, 2015. https://www.bbc.com/news/technology-32704566.

BBC News. "Colombian Anti-government Protesters Topple Columbus Statue." June 29, 2021. https://www.bbc.com/news/world-latin-america-57651833.

Beech, Dave, and Mel Jordan. "Toppling Statues, Affective Publics, and the Lessons of the Black Lives Matter Movement," *Art & The Public Sphere* 10, no.1 (2021): 3–15.

Belfiore, Eleonora, and Oliver Bennett. "Determinants of Impact: Towards a Better Understanding of Encounters with the Arts." *Cultural Trends* 16, no. 3 (2007): 225–275.

Benjamin, Ruha. *Race After Technology: Abolitionist Tools for the New Jim Code*. London, UK: Polity, 2019.

Benjamin, Walter. *One Way Street*. Edited by Griel Marcus. Cambridge, UK: Belknap Press, 1985.

Bennett, Mia M., Janice K. Chen, Luis F. Alvarez León, and Colin J. Gleason. "The Politics of Pixels: A Review and Agenda for Critical Remote Sensing." *Progress in Human Geography* 46, no. 3 (2022): 729–752.

Berlant, Lauren. "The Commons: Infrastructures for Troubling Times." *Environment and Planning D: Society and Space* 34, no. 3 (2016): 393–419.

Berlant, Lauren. *Cruel Optimism*. Durham, NC: Duke University Press, 2011.

Berlant, Lauren. "The Face of America and the State of Emergency." *The Queen of America Goes to Washington City: Essays on Sex and Citizenship*. Durham, NC: Duke University Press, 1997.

Bertin, Jacques. *Semiology of Graphics: Diagrams, Networks, Maps*, 1967 French edition, translated by W.J. Berg. Madison: University of Wisconsin Press, 1983.

Bethel, Bradley. "The Fruit Belt Demonstrates the Importance of Neighborhood Identity." *Buffalo Rising*, April 26, 2019. https://www.buffalorising.com/2019/04/the-fruit-belt-demonstrates-the-importance-of-neighborhood-identity/.

Bhandari, Aparajita, and Rebecca Noone. "Support Local: Google Maps' Local Guides Platform, Spatial Power and Constructions of 'the Local.'" *Communication, Culture and Critique* 16, no. 3 (September 2023): 198–207. https://doi.org/10.1093/ccc/tcad018.

Bhattacharyya, Gargi. *Rethinking Racial Capitalism: Questions of Reproduction and Survival*. London, UK: Rowman and Littlefield International, 2018.

Bishop, Amanda. "A Simple and Accurate Address for Your Home Using Plus Codes," *Google India Blog*. January 27, 2022. https://blog.google/intl/en-in/products/explore-communicate/simple-and-accurate-address-your-home-using-plus-codes/.

Blight, Susan and Hayden King. *Ogimaa Mikana: Reclaiming Renaming*. Public Space Interventions initiated by Susan Blight and Hayden King under the collective *Ogimaa Mikana,* projects based in Tkaronto/Toronto, Canada. https://ogimaamikana.tumblr.com/. Accessed June 10, 2021.

Block, Sheila, and Trish Hennessy. *"'Sharing Economy' or On-Demand Service Economy? A Survey of Workers and Consumers in the Greater Toronto Area."* Canadian Centre for Policy Alternatives, 2017. https://www.policyalternatives.ca/sites/default/ files/uploads/publications/Ontario%20Office/2017/04/CCPA-ON%20sharing%20economy%20in%20the%20GTA.pdf.

Bondi, Liz and Mona. Domosh. "Other Figures in Other Places: On Feminism, Postmodernism and Geography." *Environment and Planning D: Society and Space* 10, no. 2 (1992): 199–213. 10.1068/d100199.

Bray, Hiawatha. *You Are Here: From the Compass to GPS, the History and Future of How We Find Ourselves*. New York: Basic Books, 2014.

Brock, André. *Distributed Blackness: African American Cybercultures*. New York: NYU Press, 2020.

Brouwn, Stanley. *this way brouwn*. Situational, durational art project based on map-making in Amsterdam, NL. Moma Collection. 1961.

Browne, Simone. *Dark Matters: On the Surveillance of Blackness*. Durham, NC: Duke University Press, 2015, 10.1215/9780822375302.

Bucher, Taina. "The Algorithmic Imaginary." *Information, Communication & Society* 20, no. 1 (February 2017): 30–44. 10.1080/1369118X.2016.1154086.

Buckley, Eileen. "What's in a Name? Fruit Belt Residents Say It's All About Community." *WKBW News*, April 4, 2019. https://www.wkbw.com/news/local-news/whats-in-a-name-fruit-belt-residents-say-its-all-about-community.

Buffalo Niagara Medical Campus. "Entrepreneurship." 2022. https://bnmc.org/.

Business Wire. "BreezoMeter to Uncover Insights Around Exposome and Skin." December 2, 2021. https://www.businesswire.com/news/home/20211202005666/en/L%E2%80%99Or%C3%A9al-Enters-Strategic-Partnership-With-Climate-Tech-Company-BreezoMeter-to-Uncover-Insights-Around-Exposome-and-Skin.

Cadogan, Garnette. "Black and Blue." In *The Fire This Time: A New Generation Speaks about Race*, edited by Jesmyn Ward, 129–144. New York: Scribner, 2016.

Caldwell, Georgina. "L'Oréal Signs Strategic Partnership with Breezometer to Explore the Exposome and Skin." *Global Cosmetics News*, December 6, 2021. https://www.globalcosmeticsnews.com/loreal-signs-strategic-partnership-with-breezometer-to-explore-the-exposome-and-skin.

Cardiff, Janet, and George Bures Miller. "Night Walk for Edinburgh." Video walk, 55:00. Edited by Fiona Bradley. Edinburgh: The Fruitmarket Gallery, 2019. https://cardiffmiller.com/walks/night-walk-for-edinburgh/.

Carey, Bjorn. "Stanford Engineer Bradford Parkinson, the 'Father of GPS,' Wins Prestigious Marconi Prize." *Stanford News*, May 16, 2016. https://news.stanford.edu/2016/05/16/stanford-engineer-bradford-parkinson-father-gps-wins-prestigious-marconi-prize/.

Carruth, Allison. "The Digital Cloud and the Micropolitics of Energy." *Public Culture* 26, no. 2 (2014): 339–364.

Carta Capital. "United, Favela Would Form Brazil's 5th Largest State." *RioOnWatch: Community Reporting on Rio*, February 23, 2013. Translated by Arianne Reis. https://rioonwatch.org/?p=6913.

Casey, Edward S. *Getting Back into Place: Toward a Renewed Understanding of the Place-World*. Bloomington: Indiana University Press, 2009.

Ceruzzi, Paul E. *GPS*. Cambridge, MA: MIT Press, 2018.

Chester, Tim. "How Google Put Rio's Favelas on the Map in Time for the Olympics." *Mashable*, August 5, 2016. https://mashable.com/article/google-maps-rio-favelas.

Cheung, Danny. "Mapping Stories With a New Street View Trekker." *The Keyword (blog)*. Google. December 18, 2018. https://blog.google/products/maps/mapping-stories-new-street-view-trekker/.

Chow, Rey. *The Age of the World Target: Self-Referentiality in War, Theory and Comparative Work*. Durham, NC: Duke University Press, 2006.

Chowdhry, Amit. "A History of Google Acquisitions and Where They are Today." *Pulse 2*, October 4, 2008. https://pulse2.com/a-history-of-google-acquisitions-and-where-they-are-today/.

Christen, Kimberly. "Does Information Really Want to Be Free? Indigenous Knowledge Systems and the Question of Openness." *International Journal of Communication* 6 (2012): 2870–2893.

Chun, Wendy Hui Kyong. *Updating to Remain the Same: Habitual New Media*. Cambridge, MA: MIT Press, 2017.

Client Server News. "Google Snaps Up Keyhole." November 1, 2004. https://link.gale.com/apps/doc/A125837696/ITOF?u=ucl_ttda&sid=bookmark-ITOF&xid=ae412faf.

Cole, Harrison. "Tactile Cartography in the Digital Age: A Review and Research Agenda." *Progress in Human Geography* 45, no. 4 (2021): 834–854. 10.1177/0309132521995877.

Conde, Dan. "Google Touts Its Cloud Network." October 4, 2016. Archived August 16, 2017, at the Wayback Machine: https://web.archive.org/web/20170816075537/https://www.esg-global.com/blog/google-horizon2016.

Cortright, Joe. "Where Does Houston Rank Among America's Least (and Most) Segregated Cities?" *Urban Edge, Kinder Institute for Urban Research*, Rice University, September 3, 2020. https://kinder.rice.edu/urbanedge/where-does-houston-rank-among-americas-least-and-most-segregated-cities.

Cosgrove, Dennis. *Apollo's Eye: A Cartographic Genealogy of the Earth in the Western Imagination*. Baltimore, MD: Johns Hopkins University Press, 2001.

Couldry, Nick, and Ulises A. Mejias. *The Costs of Connection: How Data Colonizes Human Life and Appropriates It for Capitalism*. Stanford: Stanford University Press, 2019.

Cowan, T.L. and Jas Rault. Heavy Processing Part 1—Lesbian Processing. Heavy Processing for Digital Material (More Than a Feeling) Digital Research Ethics Collaboratory, 2020. http://www.drecollab.org/heavy-processing/

Cowan, T.L., K. Surkan and Wexler Laura. Situated Knowledges Map, FEMTECHNET. n.d. https://www.femtechnet.org/docc/feminist-mapping/situated-knowledge-map/.

Cowen, Deborah. *The Deadly Life of Logistics: Mapping Violence in Global Trade.* Minneapolis, MN: University of Minnesota Press, 2014.

Cowley, Stacy. "Google Snaps Up Digital Mapping Company: Search Engine Company Gains Database of Images of Geographic Locations." *InfoWorld.* (October 27, 2004). https://www.infoworld.com/article/2681936/google-snaps-up-digital-mapping-company.html

Crawford, Kate. *Atlas of AI: Power, Politics, and the Planetary Costs of Artificial Intelligence.* Yale University Press, 2021.

Crawford, Kate and Vladan Joler. "Anatomy of an AI System: The Amazon Echo as an Anatomical Map of Human Labor, Data and Planetary Resources." *Diagram. AI Now Institute and Share Lab*, September 7, 2018. https://anatomyof.ai/.

Cuthbertson, Anthon. "Google Maps Error Sees Wrong House Demolished." *Newsweek*, March 24, 2016. https://www.newsweek.com/google-maps-error-sees-wrong-house-demolished-mistake-440256.

Daniel, Miriam. "Immersive View Coming Soon to Maps—Plus More Updates." *The Keyword (blog).* Google, May 11, 2022. https://blog.google/products/maps/three-maps-updates-io-2022/.

Data Centre Frontier. "Why Chicago Is a Geostrategic Destination for Data Center Investment," *Voices of the Industry.* June 30, 2021. https://datacenterfrontier.com/why-chicago-is-a-geostrategic-destination-for-data-center-investment/.

de Souza e Silva, Adriana, and Jordan Frith. "Locative Mobile Social Networks: Mapping Communication and Location in Urban Spaces." *Mobilities* 5, no. 4 (November 2010): 485–505.

de Souza e Silva Adriana, and Jordan Frith. *Mobile Interfaces in Public Spaces: Locational Privacy, Control, and Urban Sociability.* New York, NY: Routledge, 2012.

Debord, Guy. "One Step Back." Edited and translated by Tim McDonough. In *Guy Debord and the Situationists International: Texts and Documents*, 25–28. Cambridge, MA: MIT Press, 2002.

Debord, Guy. "Report on the Construction of Situations and on the Terms of Organization and Action of the International Situationist Tendency." Edited and translated by Tim McDonough. In *Guy Debord and the Situationists International: Texts and Documents*, 29–50. Cambridge, MA: MIT Press, 2002.

Dewey, Caitlin. "Fruit Belt Fights for Its Name Over Fears Big Tech Is Erasing It." *The Buffalo News*, March 17, 2019. https://buffalonews.com/2019/03/17/fruitbeltfights-for-its-name-over-fears-big-tech-is-erasing-it/.

Dickenson, Courtney. "Protesters Toss Statue of Explorer James Cook into Victoria Harbour." *CBC News*, July 1, 2021. https://www.cbc.ca/news/canada/british-columbia/victoria-captain-cook-statue-vandalized-1.6088828.

Dicker, Russel. "3 New Ways to Navigate More Sustainably with Maps." *The Keyword (blog).* Google. October 6, 2021. https://blog.google/products/maps/3-new-ways-navigate-more-sustainably-maps/.

Dicker, Russel. "All the Ways Google Gets Street View Images." June 20, 2022. *Wired Magzine video*, 9:39. https://www.wired.com/video/watch/wired-news-and-science-google-street-view.

do Couto, Bruno Gontyjo. "Ideology and Utopia of Brasilia: Disputes Over Design in Modern Brazil." *Sociedade e estado* 28, no. 3 (2013): 730–731.

Doig, Will. "How Do We Get More Power?" *Open Society Foundations*, May 4, 2020. https://www.opensocietyfoundations.org/voices/how-do-we-get-more-power/ episode/fruit-belt.

Drucker, Johanna. *Visualization and Interpretation: Humanistic Approaches to Display*, 1st ed. Cambridge, MA: MIT Press, 2020.

Drucker, Johanna. *Graphesis: Visual Forms of Knowledge Production*. Cambridge, MA: Harvard University Press, 2014.

Duffy, Brooke Erin. *(Not) Getting Paid to do What You Love*. New Haven, CT: Yale University Press, 2016.

Eades, Gwilym L. *Maps and Memes: Redrawing Culture, Place, and Identity in Indigenous Communities*. Montreal: McGill-Queen's University Press, 2015.

Edney, Matthew. "Some Thoughts on Jacques Bertin's Cartographic Semiology." *Mapping As Process (blog)*. August 20, 2021. https://www.mappingasprocess.net/ blog/2021/8/20/some-thoughts-on-jacques-bertins-cartographic-semiology.

Eleyan, Derar, Abed Othman, and Amna Eleyan. "Enhancing Software Comments Readability Using Flesch Reading Ease Score." *Information* 11, no. 9 (2020): 430. 10.3390/info11090430.

El Saadawi, Nawal. "Dissidence and Creativity." *Women: A Cultural Review* 6, no. 1 (1995): 1–17.

Elwood, Sarah. "Digital Geographies, Feminist Relationality, Black and Queer Code Studies: Thriving Otherwise." *Progress in Human Geography* 45, no. 2 (2021): 209–228.

Elwood, Sarah. "Critical Issues in Participatory Gis: Deconstructions, Reconstructions, and New Research Directions." *Transactions in GIS* 10, no. 5 (2006): 693–708.

Elwood, Sarah, and Agnieszka Leszczynski. "Feminist digital geographies." *Gender, Place & Culture* 25, no. 5 (2018): 629–644.

Elwood, Sarah, and Agnieszka Leszczynski. "New Spatial Media, New Knowledge Politics," *Transactions of the Institute of British Geographers*. 38 (2013): 544–559.

Epstein, Johnathan D. "Medical Campus Development Boom Starts to Push into Fruit Belt." *Buffalo News*, September 13, 2019. https://buffalonews.com/news/local/ medical-campus-development-boom-starts-to-push-into-fruit-belt/article_42eaee12-a0ef-5057-93d8-d30ffaee6e7f.html.

Eubanks, Virginia. *Digital Dead End: Fighting for Social Justice in the Information Age*. Cambridge, MA: MIT Press, 2011.

Farman, Jason. "Map Interfaces and Production of Locative Media Space." In *Locative Media*, edited by Rowin Wilken and Gerard Goggin, 83–93. London: Routledge, 2015.

Farman, Jason. "Mapping the Digital Empire: Google Earth and the Process of Postmodern Cartography." *New Media & Society* 12, no. 6 (2010): 869–888. 10.1177/ 1461444809350900.

Farman, Jason. *Mobile Interface Theory*. New York: Routledge, 2011.

Fast, Karin, and Pablo Abend. "Introduction to Geomedia Histories." *New Media & Society* 24, no. 11 (October 2022): 2385–2395. 10.1177/14614448221122168.

Filliou, Robert. *Teaching and Learning as Performance Arts*. Cologne: Verlag Gebrl König, 1979.

Fiske, John. "Surveilling the City: Whiteness, the Black Man and Democratic Totalitarianism." *Theory, Culture & Society* 15, no. 2 (May 1998): 67–88. 10.1177/ 026327698015002003.

Flesch, R. "Reply to Simplification of Flesch Reading Ease Formula." *Journal of Applied Psychology* 36 (February 1952): 54–55.

Frampton, Adam, Clara Wong, and Johnathan D. Solomon. *Cities Without Ground: A Hong Kong Guidebook*. Navato: Oro Editions, 2012.

Frith, Jordan. *Smartphones as Locative Media*. London, UK: Polity Press, 2015.

Frith, Jordan, and Didem Özkul. "Mobile Media Beyond Mobile Phones." *Mobile Media & Communication* 7, no. 3 (September 2019): 293–302.

Fruit Belt Community Land Trust. "Fruit Belt Community Land Trust: Development Without Displacement." Archived by Wayback Machine June 20, 2021. https://web.archive.org/web/20210620032326/https://fruitbelt-clt.org/

Galloway, Alexander. *The Interface Effect*. Cambridge, UK: Polity, 2012.

Galloway, Anne, and Matthew Ward. "Locative Media as Socialising and Spatializing Practice: Learning from Archaeology." *Leonardo Electronic Almanac* 14, no. 3 (July 2006).

Gangadharan, Seeta Peña. "The Downside of Digital Inclusion: Expectations and Experiences of Privacy and Surveillance Among Marginal Internet Users." *New Media & Society* 19, no. 4 (November 9, 2015): 597–615. 10.1177/1461444815614053.

Gannes, Liz. "Ten Years of Google Maps, From Slashdot to Ground Truth: Ten episodes from the dawning days of Google Maps," Vox, Febrary 8, 2015. https://www.vox.com/2015/2/8/11558788/ten-years-of-google-maps-from-slashdot-to-ground-truth

Garzón, Catalina, Heather Cooley, Matthew Heberger, Eli Moore, Lucy Allen, Eyal Matalon, Anna Doty, and the Oakland Climate Action Coalition. *Community Based Climate Adaptation Planning: Case Study of Oakland, California*. White paper prepared by Pacific Institute for California Energy Commission. July 2012. https://woeip.org/wp-content/uploads/2020/11/WOEIP-research-community-based-climate-planning-Oakland.pdf.

Gaspar, Joaquim Alves. "Revisiting the Mercator World Map of 1569: an Assessment of Navigational Accuracy." *The Journal of Navigation* 69, no. 6 (2016): 1183–1196.

Gentzel, Peter, Jeffrey Wimmer, and Ruben Schlagowski. "Doing Google Maps: Everyday Use and the Image of Space in a Surveillance Capitalism Centrepiece." *Digital Culture and Society* 7, no. 2 (2021): 159–184.

Gillespie, Tarleton. *Custodians of the Internet: Platforms, Content Moderation, and the Hidden Decisions That Shape Social Media*. New Haven: Yale University Press, 2018.

Gilliard, Chris. "Friction-Free Racism." *Real Life*. (October 15, 2018). https://reallifemag.com/friction-free-racism/. Accessed 17 April 2023.

Gilliard, Chris. "From Redlining to Digital Redlining." Academic Technology Expo 2018. The University of Oklahoma. YouTube video. January 31, 2018. https://www.youtube.com/watch?v=MEPI7YctRqY&t=313s.

Gilliard, Chris. "Pedagogy and the Logic of Platforms." *EDUCAUSE Review* 52, no. 4 (July/August 2017). https://er.educause.edu/articles/2017/7/pedagogy-and-the-logic-of-platforms.

Gilliard, Chris, and Hugh Culik. "Digital Redlining, Access, and Privacy." Common Sense Education. May 24, 2016. https://www.commonsense.org/education/articles/digital-redlining-access-and-privacy.

Gilmore, James N., and Bailey Troutman. "Articulating Infrastructure to Water: Agri-culture and Google's South Carolina Data Center." *International Journal of Cultural Studies* 23, no. 6 (2020): 916–931.

Gilmore, Michael T. *Surface and Depth: The Quest for Legibility in American Culture*. Oxford: Oxford University Press, 2003.

Gilmore, Ruth Wilson. "Fatal Couplings of Power and Difference: Notes on Racism and Geography." *The Professional Geographer* 54, no. 1 (March 15, 2010): 15–24.

Glük, Louise. "Prism I." In *Averno*. Penguin, 2006.

Goggin, Gerard, and Larissa Hjorth. "The Question of Mobile Media." In *Mobile Technologies: From Telecommunications to Media*, edited by Gerald Goggin and Larissa Hjorth, 3–8. New York: Routledge, 2009.

Gonzalez, Priscilla A., Meredith Minkler, Analilia P. Garcia, Margaret Gordon, Catalina Garzón, Meena Palaniappan, Swati Prakash, and Brian Beveridge. "Community-Based Participatory Research and Policy Advocacy to Reduce Diesel Exposure in West Oakland, California." *American Journal of Public Health* 101, no. S1 (2011): S166–S175. https://www.proquest.com/docview/906290394.

Gonzalez, Rosa. *Community-Driven Climate Resilience Planning: A Framework, Version 2.0*. Edited by Taj James and Jovida Ross. National Association of Climate Resilience Planners. May 2017. https://kresge.org/sites/default/files/library/community_drive_resilience_planning_from_movement_strategy_center.pdf.

Goodchild, Michael F., and Donald G. Janelle, eds. *Spatially Integrated Social Science*. New York: Oxford Press, 2004.

Google. "Air Quality." Earth Outreach. Accessed May 30, 2023. https://www.google.com/earth/outreach/special-projects/air-quality/.

Google. "Beyond the Map, Rio de Janeiro—Luis' Story." YouTube video, 3:46. July 31, 2016. https://www.youtube.com/watch?v=6Q9Yuayb2J4.

Google. "Beyond the Map, Rio de Janeiro—Mapping the Favelas." YouTube Video, 1:32. August 1, 2016. https://www.youtube.com/watch?v=MpgDIq_veLE.

Google. "Beyond the Map, Rio de Janeiro—Paloma's Story." YouTube Video, 2:30. July 31, 2016. https://www.youtube.com/watch?v=l-oG47usLdI.

Google. "Beyond the Map, Rio de Janeiro—Ricardo's Story." YouTube Video, 2:52. July 31, 2016. https://www.youtube.com/watch?v=lRhqsxMOZHs.

Google. "BreezoMeter: Delivering Global Environmental Information with Google Cloud." Google Cloud. https://cloud.google.com/customers/breezometer/.

Google. "Environmental Insights Explorer." Environmental Insights Explorer. Accessed May 30, 2023. https://insights.sustainability.google/labs/airquality.

Google. "Google Data Centre." Accessed May 20, 2023. https://www.google.com/about/datacenters/locations/hamina/

Google. "Google Maps/Google Earth Additional Terms of Service." Accessed March 5, 2021. https://maps.google.com/help/terms_maps/

Google. "Keyhole Markup Language." Google for Developers. Accessed March 5, 2021. https://developers.google.com/kml/

Google. "Rio: Beyond the Map with Google." Google Arts & Culture. https://artsandculture.google.com/project/rio-de-janeiro.

Google Cloud Tech. "Horizon – Jen Fitzpatrick – Going Beyond the Map." Presented by Jen Fitzpatrick. October 17, 2016, YouTube video, 14:57. https://www.youtube.com/watch?v=bPo_fLznqZ0

Google for Developers. "Keynote (Google I/O '18)." May 18, 2018, YouTube video, 1:46:31. https://www.youtube.com/watch?v=ogfYd705cRs.

Google for Developers. "Making the World Your Own with Google Maps APIs (Google I/O '17)." Presented by Joël Kalmanowicz. YouTube video, 26:02. May 19, 2017. https://www.youtube.com/watch?v=vLRutvtJwLg.

Google Maps. "Addresses for Everyone." https://maps.google.com/pluscodes/learn/#case-studies.

Google Maps. "Bringing Your Map to Life, One Image at a Time." Accessed October 10, 2023. https://www.google.com/intl/ALL/streetview/.

Google Maps. "Explore and Navigate Your World." Accessed March 5, 2021. https://www.google.com/maps/about/#!/.

Google Maps. "Explore and Navigate Your World." Accessed March 5, 2021. https://www.google.com/maps/about/#!/.

Google Maps Help. "Find Wheelchair-Accessible Places." Accessed October 13, 2023. https://support.google.com/maps/answer/9882117.

Google Maps. "Google Maps: There's More to Explore." YouTube, 2013 (original video is no longer available, but can be viewed on the animator's website at https://lumichang.work/google-maps).

Google Maps. "Learn about Plus Codes." Accessed June 10, 2022. https://maps.google.com/pluscodes/learn/.

Google Maps. "My Maps." Accessed May 5, 2021. https://www.google.com/maps/about/mymaps/.

Google Maps. "Plus Codes." https://maps.google.com/pluscodes/.

Google Maps. "Why We Map the World." Presented by Luiz André Barroso. YouTube video, 3:37. December 19, 2018. https://www.youtube.com/watch?v=TiK3i37HPyM.

Google Search. "Our Approach to Search." Accessed October 10, 2023. https://www.google.com/search/howsearchworks/our-approach/.

Gordon, Eric, and Adriana de Souza e Silva. *Net Locality: Why Location Matters in a Networked World*, 1st ed. Hoboken, NJ: Wiley-Blackwell, 2011.

Graham, Mark, and Martin Dittus. *Geographies of Digital Exclusions: Data and Inequity*. London: Pluto Press, 2022.

Graham, Steve, and Simon Marvin. *Splintering Urbanism: Networked Infrastructures, Technological Mobilities and the Urban Condition*. London, UK: Routledge, 2001.

Grande, Sandy. *Red Pedagogy: Native American Social and Political Thought*. Lanham, MD: Rowman & Littlefield Publishers, 2004.

Gray, Mary, and Siddharth Suri. *Ghost Work: How to Stop Silicon Valley from Building a New Global Underclass*. Boston, MA: Harcourt Publishing, 2019.

Gudeman, Stephen. "The New Captains of Information." *Anthropology Today* 14, no. 1 (February 1, 1998): 1–3.

Halegoua, Germaine R. *The Digital City: Media and the Social Production of Place*. New York, NY: NYU Press, 2019.

Hamraie, Aimi. "Mapping Access: Digital Humanities, Disability Justice, and Sociospatial Practice." *American Quarterly* 70, no. 3 (2018): 455–482.

Hamraie, Aimi. *Building Access: Universal Design and the Politics of Disability*. University of Minnesota Press, 2017.

Haraway, Donna. "Situated Knowledges: The Science Question in Feminism and the Privilege of Partial Perspective." *Feminist Studies* 14, no. 3 (1988): 575–599.

Harley, John Brian. "Can There Be a Cartographic Ethics?" *Cartographic Perspectives*, 10 (1991): 9–16. 10.14714/CP10.1053

Harley, John Brian and Paul Laxton. *The New Nature of Maps: Essays in the History of Cartography*. Baltimore, MA: Johns Hopkins University Press, 2002.

Harris, Dianne. *Little White Houses: How the Postwar Home Constructed Race in America*. Minneapolis, MN: University of Minnesota Press, 2012.

Harris, Michael. *The End of Absence: Reclaiming What We've Lost in a World of Constant Connection*. London, UK: Portfolio Penguin, 2014.

Harvey, David. "The Spatial Fix: Hegel, von Thünen and Marx." *Antipode* 13, no. 3 (December 1981): 1–12. 10.1111/j.1467-8330.1981.tb00312.x.

Harvey, David. *Paris, Capital of Modernity*. New York: Routledge, 2003. 10.4324/9780203508619.

Hawkins, Andrew J. "The Deep Dive into Google Maps Is Fascinating." *The Verge*, December 24, 2017. https://www.theverge.com/2017/12/24/16801334/google-maps-justin-obeirne-cartographer-apple-waymo.

Hays, Jeremy, Clara Landeiro, and Jane Rongerude. *Neighborhood Knowledge for Change: The West Oakland Environmental Indicators Project*. Project directors Steve Costa, Meena Palaniappan, and Arlene K. Wong. Oakland, CA: Pacific Institute for Studies in Development, Environment, and Security, 2002. https://woeip.org/wp-content/uploads/2020/11/WOEIP-research-neighborhood_knowledge_for_change3.pdf.

Hern, Alex. "Google Maps Hides an Image of the Android Robot Urinating on Apple." *The Guardian*, April 24, 2015. https://www.theguardian.com/technology/2015/apr/24/google-maps-hides-an-image-of-the-android-robot-pissing-on-apple.

Hesselberth, Pepetia, Janna Houwen, Esther Peeren, and Ruby de Vos. "Introduction." In *Legibility in the Age of Signs and Machines, 1–18*. Leiden: Brill, 2018. 10.1163/9789004376175_002.

Hicks, Alison. "Risky (Information) Business: An Informational Risk Research Agenda." *Journal of Documentation*, 79, no. 5 (2023): 1147–1163.

Highmore, Ben. "Georges Perec and the Significance of the Insignificant." In *The Afterlives of Georges Perec*, edited by Rowan Wilken and Justin Clemens, 105–119. Edinburgh University Press, 2017. 10.3366/edinburgh/9781474401241.003.0012.

Hockenberry, Matthew, Nicole Starosielski, and Susan Zieger. *Assembly Codes: The Logistics of Media*. Durham, NC: Duke University Press, 2021. 10.1215/9781478013037.

Hoffmann, Anna Lauren. "Terms of Inclusion: Data, Discourse, Violence." *New Media & Society* 23, no. 12 (2021): 3539–3556. 10.1177/1461444820958725.

Hoffmann, Anna Lauren, and Anne Jonas. "Recasting Justice for Internet and Online Industry Research Ethics." In *Internet Research Ethics for the Social Age: New Cases and Challenges*, edited by Michael Zimmer and Katharina Kinder-Kurlanda. Digital Formations Volume 108, (2016): https://papers.ssrn.com/abstract=2836690.

Hogan, Mél. "Big Data Ecologies." *Ephemera* 18, no. 3 (2018): 631–657.

Hogan, Mél. "Data Flows and Water Woes: The Utah Data Centre." *Big Data & Society* 2, no. 2 (2015). 10.1177/2053951715592429.

Hogan, Mél. "Facebook Data Storage Centers as the Archive's Underbelly." *Television & New Media* 16, no. 1 (2015): 3–18.

hooks, bell. "Choosing the Margin as a Space of Radical Openness." *Framework: The Journal of Cinema and Media*, no. 36 (1989): 15–23. http://www.jstor.org/stable/44111660.

Hough, Andrew. "Lauren Rosenberg: US Woman Sues Google 'After Maps Directions Caused Accident.'" *The Telegraph*, June 2, 2010. https://www.telegraph.co.uk/technology/google/7795460/Lauren-Rosenberg-US-woman-sues-Google-after-Maps-directions-caused-accident.html.

Hu, Tung-Hui. *A Prehistory of the Cloud*. Cambridge: MA: MIT Press, 2016.

Introna, Lucas D., and Helen Nissenbaum. "Shaping the Web: Why the Politics of Search Engines Matters." *The Information Society* 16, no. 3 (2000): 169–185. 10.1080/01972240050133634.

Investopedia. "How Does Google Maps Make Money?" Reviewed by Charlene Rhineheart. September 11, 2022. https://www.investopedia.com/articles/investing/061115/how-does-google-maps-makes-money.asp.

Irani, Lilly. *Chasing Innovation: Making Entrepreneurial Citizens in Modern India*. Princeton, NJ: Princeton University Press, 2019.

Jacobson, Kate, and Mél Hogan. "Retrofitted Data Centres: A New World in the Shell of the Old." *Work Organization, Labour and Globalization* 13, no. 2 (Winter 2019): 78–94.

John Muir Trust. "Online Routes Could Put Walkers at Risk." July 16, 2021. https://www.johnmuirtrust.org/whats-new/news/929-online-routes-could-put-walkers-at-risk.

Junius, Lie. "New Data on Data Centers: How Google Helps Regions Grow." *The Keyword (blog)*. Google. February 20, 2018. https://blog.google/around-the-globe/google-europe/new-data-data-centers-how-google-helps-regions-grow/.

Justie, Brian. "Little History of CAPTCHA." *Internet Histories* 5, no. 1 (2021): 30–47.

Kadenge, Maxwell, and Dion Nkomo. "The Politics of the English Language in Zimbabwe." *Language Matters* 42, no. 2 (October 27, 2010): 248–263. 10.1080/10228195.2011.581679.

Kaplan, Caren. "Precision Targets: GPS and the Militarization of US Consumer Identity." *American Quarterly* 58, no. 3 (2006): 693–714.

Kassens-Noor, Eva, Chirstopher Gaffney, Joe Messina, and Eric Phillips. "Olympic Transport Legacies: Rio de Janeiro's Bus Rapid Transit System." *Journal of Planning Education and Research* 38, no. 1 (December 2016): 13–24. 10.1177/073 9456×16683228.

Kassens-Noor, Eva, Mark Wilson, Sven Müller, Brij Maharaj, and Laura Huntoon. "Towards a Mega-Event Legacy Framework." *Leisure Studies* 34, no. 6 (April 29, 2015): 665–671. 10.1080/02614367.2015.1035316.

Kastrenaskes, Jacob. "Do Not Blame Google Maps When You Tear Down the Wrong House." *The Verge*, March 25, 2016. https://www.theverge.com/2016/3/25/11306410/google-maps-error-leads-demolition-company-to-wrong-home.

Kennedy, Merrit. "Google Maps Leads About 100 Drivers into a 'Muddy Mess' in Colorado." *NPR News*, June 27, 2019. https://www.npr.org/2019/06/27/736572732/google-maps-leads-about-100-drivers-into-a-muddy-mess-in-colorado.

Kinder Institute for Urban Research. *Houston Disparity Atlas*. Houston, TX: Rice University, March 1, 2016. https://kinder.rice.edu/research/houston-disparity-atlas.

Kinder Institute for Urban Research. *Houston Region Diversity Report*. Houston, TX: Rice University, March 1, 2012. https://kinder.rice.edu/research/houston-region-diversity-report.

Kingsbury, James A. *Groundwater Quality in the Surficial Aquifer System, Southeastern United States*. Report for National Water Quality Program, National Water-Quality Assessment Project. July 22, 2022. https://pubs.usgs.gov/publication/fs20223035/full.

Kirby, Emma, Ash Watson, Brendan Churchill, Brady Robards, and Lucas LaRochelle. "Queering the Map: Stories of Love, Loss and (be)Longing within a Digital Cartographic Archive." *Media, Culture & Society*, 43, no. 6 (2021): 1043–1060. 10.1177/0163443720986005.

Kitchin, Rob. "The Real-Time City? Big Data and Smart Urbanism." *GeoJournal* 79, no. 1 (2014):1–14. 10.1007/s10708-013-9516-8.

Kitchin, Rob, and Martin Dodge. *Code/Space: Software and Everyday Life*. MIT Press, 2014.

Kitchin, Rob, Martin Dodge, and Chris Perkins. "Introductory Essay: Conceptualising Mapping." In *The Map Reader*, edited by Rob Kitchin, Martin Dodge, and Chris Perkins, xix–xxiii. Oxford: Wiley-Blackwell Press, 2011.

Knight, Erin. "2021 Report Shows Canada's Cell Phone Prices STILL Among Most Expensive Globally." *Open Media*, October 14, 2021. https://openmedia.org/article/item/2021-rewheel-report-shows-canadas-cell-phone-prices-still-among-most-expensive-globally.

Kuo, Rachel, and Alice Marwick. "Critical Disinformation Studies." *Harvard Kennedy School (HKS) Misinformation Review* 2, no. 4 (2021). https://misinforeview.hks.harvard.edu/article/critical-disinformation-studies-history-power-and-politics/.

Kwan, Mei-Po. "Algorithmic Geographies: Big Data, Algorithmic Uncertainty, and the Production of Geographic Knowledge." *Annals of the American Association of Geographers* 106, no. 2 (2016): 274–282.

Kwan, Mei-Po. "Feminist Visualization: Re-Envisioning GIS as a Method in Feminist Geographic Research." *Annals of the Association of American Geographers* 92, no. 4 (March 2010): 645–661. 10.1111/14678306.00309 10.1111/1467-8306.00309.

Kwon, Miwon. *One Place after Another: Site-Specific Art and Locational Identity*. Cambridge, MA: MIT Press, 2002.

Larkin, Brian. "The Politics and Poetics of Infrastructure." *Annual Review of Anthropology* 42 (2013): 327–343.

Lee, Jiyeong. "Global Positioning/GPS." In *International Encyclopedia on Human Geography*, edited by Rob Kitchin and Nigel Thrift, 548–555. New York, NY: Elsevier Science, 2009.

Lee, Micky. "A Political Economic Critique of Google Maps and Google Earth." *Information, Communication and Society* 13, no. 6 (2010): 909–928.

Lee, Su-Ying. "Reclaiming/Renaming." *Institute for Public Art*, 2021. https://www. instituteforpublicart.org/case-studies/reclaiming-renaming/.

Lehmann, Claire. "Stanley Brouwn." In *Artists Who Make Books*, edited by Andrew Roth, Phillip Aarons, Claire Lehmann, and Jeffrey Kastner, 55–60. London, UK: Phaidon, 2017.

Leszczynski, Agnieszka. "Situating the Geoweb in Political Economy." *Progress in Human Geography*, 36, no. 1 (2012): 72–89.

Leopkey, Becca, and Milena Parent. "The Governance of Olympic Legacy: Process, Actors and Mechanisms." *Leisure Studies* 36, no. 3 (2017): 1–14. 10.1080/02614367. 2016.1141973.

Li, Mark, and Zhoul Bailiang. "Discover the Action Around You with the Updated Google Maps." *The Keyword (blog)*. Google. July 25, 2016. https://blog.google/ products/maps/discover-action-around-you-with-updated/.

Liboiron, Max. *Pollution Is Colonialism*. Durham, NC: Duke University Press, 2021.

Licoppe, Christian. "'An Attempt at Exhausting an Augmented Place in Paris': Georges Perec, Observer-Writer of Urban Life, as a Mobile Locative Media User." In *The Afterlives of Georges Perec*, edited by Rowan Wilken and Justin Clemens, 220–225. Edinburgh, UK: Edinburgh University Press, 2017. 10.3366/edinburgh/ 9781474401241.003.0012.

Light, Jennifer. "Discriminating Appraisals: Cartography, Computation, and Access to Federal Mortgage Insurance in the 1930s." *Technology and Culture* 52, no. 3 (2011): 485–522. 10.1353/tech.2011.0111.

Livesey, Graham. "From the Infraordinary to the Extraordinary: Georges Perec and Domesticity." *Architectural Research Quarterly* 26, no. 3 (September 2022): 247–253. 10.1017/S1359135522000471.

Lloyd, Annemaree. "Framing Information Literacy as Information Practice: Site Ontology and Practice Theory." *Journal of Documentation* 66, no. 2 (2010): 245–258. 10.1108/00220411011023643.

Lloyd, Annemaree. *Information Literacy Landscapes: Information Literacy in Education, Workplace, and Everyday Contexts*. Oxford: Chandos, 2010.

Lloyd, Annemaree. "Information Literacy Landscapes: An Emerging Picture." *Journal of Documentation* 62, no. 5 (2006): 570–583.

London Transport Museum. "Transforming the Tube Map: Harry Beck's Iconic Design." https://www.ltmuseum.co.uk/collections/stories/design/transforming-tube-map-harry-becks-iconic-design. Accessed June 3, 2023.

Loukissas, Yanni Alexander. *All Data Are Local: Thinking Critically in a Data-Driven Society*. Cambridge, MA: MIT Press, 2019.

Luque-Ayala, Andrés, and Flávia Neves Maia. "Digital Territories: Google Maps as a Political Technique in the Re-Making of Urban Informality." *Environment and Planning D: Society and Space* 37, no. 3 (2019): 449–467.

Lynch, Kevin. *The Image of the City*. Cambridge, MA: MIT Press, 1960.

MacEachren, Alan M. "(re)Considering Bertin in the Age of Big Data and Visual Analytics." *Cartography and Geographic Information Science* 46, no. 2 (2019): 101–118. 10.1080/15230406.2018.1507758.

Madrigal, Alexis. "How Google Builds Its Maps and What It Means for the Future of Everything." *The Atlantic*, September 6, 2012. https://www.theatlantic.com/ technology/archive/2012/09/how-google-builds-its-maps-and-what-it-means-for-the-future-of-everything/261913/.

Magalhães, João Carlos, and Nick Couldry. "Giving by Taking Away: Big Tech, Data Colonialism, and the Reconfiguration of Social Good." *International Journal of Communication* 15 (2021): 343–362.

Malatino, Hil. *Trans Care*. Minneapolis, MN: University of Minnesota Press, 2022.

Malatino, Hil. "Future Fatigue: Trans Intimacies and Trans Presents (or How to Survive the Interregnum)." *TSQ* 6, no. 4 (November 2019): 635–658. doi: https://doi-org.ezproxy.lib.gla.ac.uk/10.1215/23289252-7771796

Malczyk, Katie. "Let Google be Your Holiday Travel Tour Guide, 2019." *The Keyword (blog)*. Google. December 13, 2019. https://www.blog.google/products/maps/let-google-be-your-holiday-travel-tour-guide/.

Mamers, Danielle Taschereau. "Settler Colonial Ways of Seeing: Documentary Governance of Indigenous Life in Canada and Its Disruption." PhD diss., University of Western Ontario, 2017.

Marcus, Greil. "The Long Walk of the Situationist International." In *Guy Debord and the Situationist International: Text and Documents*, edited by Tom McDonough, 1–19. Cambridge, MA: MIT Press, 2004.

Markoff, John. "Technology: That's the Weather, and Now, Let's Go to the Cellphone for the Traffic." *New York Times*, March 1, 2004.

Martin, Craig. "Shipping Container Mobilities, Seamless Compatibility and the Global Surface of Integration." *Environment and Planning A* 45, no. 5 (2013): 1021–1036.

Martin, Craig. "Controlling Flow: On the Logistics of Distributive Space." In *Architecture in the Space of Flows*, edited by Andrew Ballantyne and Christopher L. Smith, 147–160. London, UK: Routledge, 2012.

Martin, David. "How We're Giving Everyone, Everywhere an Address." *The Keyword (blog)*. Google. September 29, 2020. https://blog.google/products/maps/google-maps-101-giving-everyone-everywhere-an-address/.

Massey, Doreen. *World City*. Cambridge, U.K: Polity, 2007.

Massey, Doreen. *For Space*. Thousand Oaks, CA: Sage, 2005.

Massey, Doreen. *Space, Place, and Gender*. Minneapolis, MN: University of Minnesota Press, 1994.

Massey, Doreen. "A Global Sense of Place." *Marxism Today* 38 (1991): 24–29.

Matchar, Emily. "Mapping Rio's Favelas." *Smithsonian Magazine*, July 15, 2016. https://www.smithsonianmag.com/innovation/mapping-rios-favelas-180959816/.

Mattern, Shannon. *A City Is Not a Computer: Other Urban Intelligences*. Princeton: Princeton University Press, 2021.

Mattern, Shannon. "Maintenance and Care." *Places Journal* (November 2018). https://placesjournal.org/article/maintenance-and-care/.

Mattern, Shannon. "A City Is Not a Computer." *Places Journal* (February 2017). 10.22269/170207.

Mattern, Shannon. *Code and Clay, Data and Dirt: Five Thousand Years of Urban Media*. Minneapolis: University of Minnesota Press, 2017.

Mattern, Shannon. *Deep Mapping the Media City*. Minneapolis: University of Minnesota Press, 2015.

Maynard, Robyn. *Policing Black Lives: State Violence in Canada from Slavery to Present*. Halifax: Fernwood Publishing, 2017.

Maynard, Robyn, and Leanne Betasamosake Simpson. *Rehearsals for Living*. London: Haymarket, 2022.

Maynard, Robyn, and Leanne Betasamosake Simpson. "Towards Black and Indigenous Futures on Turtle Island." In *Until We Are Free: Reflections on Black Lives Matter in Canada*, edited by Rodney Diverlus, Sandy Hudson, and Syrus Marcus Ware, 75–93. Regina: University of Regina Press, 2020.

McCullough, Malcolm. "On the Urbanism of Locative Media." *Places* 18, no. 2 (August 2006): 26–29.

McKittrick, Katherine. *Demonic Grounds: Black Women and the Cartographies of Struggle*. Minneapolis: University of Minnesota Press, 2006.

McMillan Cottom, Tressie. "Where Platform Capitalism and Racial Capitalism Meet: The Sociology of Race and Racism in the Digital Society." *Journal of Race and Ethnicity* 6, no. 4 (2020): 441–449. 10.1177/2332649220949473.

McQuire, Scott. "Learning From Street View: Lessons in Urban Visuality." In *Visual and Multimodal Urban Sociology, Part A, Research in Urban Sociology, Vol. 18A*, edited by Luc Pauwels, 141–160. Bingley: Emerald Publishing Limited, 2023. 10.1108/S1047-00422023000018A006.

McQuire, Scott. "One Map to Rule Them All? Google Maps as Digital Technical Object." *Communication and the Public* 4, no. 2 (June 2019): 150–165. 10.1177/205 7047319850192.

McQuire, Scott. *Geomedia: Networked Cities and the Future of Public Space*. Cambridge, UK: Polity, 2016.

Menendian, Stephen, Samir Gambhir, and Arthur Gailes. "Twenty-First Century Racial Residential Segregation in the United States." The Roots of Structural Racism Project. Othering and Belonging Institute. Updated June 30, 2021. https://belonging.berkeley.edu/roots-structural-racism.

Mervyn, Kieran, Anoush Simon, and David K. Allen. "Digital Inclusion and Social Inclusion: A Tale of Two Cities." *Information, Communication and Society* 17, no. 9 (February 11, 2014): 1086–1104. 10.1080/1369118X.2013.877952.

Milan, Stefania, and Emiliano Treré. "Big Data from the South(s): Beyond Data Universalism." *Television & New Media* 20, no. 4 (April 11, 2019): 319–335. 10.1177/1527476419837739.

Miller, Greg. "The Huge, Unseen Operation Behind the Accuracy of Google Maps." *Wired Magazine*, December 8, 2014. https://www.wired.com/2014/12/google-maps-ground-truth/.

Milman, Oliver. "Christopher Columbus Statues Toppled in Virginia and Beheaded in Boston." *The Guardian*, June 11, 2020. https://www.theguardian.com/us-news/2020/jun/10/christopher-columbus-statue-toppled-virginia.

Milner, Greg. "Death by GPS? Are Satnav Changing Our Brains?" *The Guardian*, June 25, 2016. https://www.theguardian.com/technology/2016/jun/25/gps-horror-stories-driving-satnav-greg-milner.

Mirzoeff, Nicholas. *The Right to Look: A Counterhistory of Visuality*. Durham, NC: Duke University Press, 2011.

Mosendz, Polly. "Google Limits Access to Map Maker Tool After Peeing Android Mascot Prank," *Newsweek*, May 15, 2015. https://www.newsweek.com/google-limits-access-map-maker-tool-after-peeing-android-mascot-prank-331144.

Mountaineering Scotland. "Google Maps Risk on Ben Nevis." July 15, 2021. https://www.mountaineering.scot/news/google-maps-risk-on-ben-nevis.

Mpofu, Phillip, and Abiodun Salawu. "African Language Use in the Digital Public Sphere: Functionality of the Localised Google Webpage in Zimbabwe." *South African Journal of African Languages* 40, no. 1 (April 1, 2020): 76–84. 10.1080/02572117.2020.1733833.

Muehlenhaus, Ian. *Web Cartography: Map Design for Interactive and Mobile Devices*, 1st edition. Boca Raton, FL: CRC Press, 2014.

Mulvin, Dylan. *Proxies: The Cultural Work of Standing In*. Cambridge, MA: MIT Press, 2021.

Myers, Alex. "Bradford Parkinson: Hero of GPS." *Stanford Engineering Magazine, School News*, June 8, 2012. https://engineering.stanford.edu/magazine/bradford-parkinson-hero-gps.

Naim, Oren. "Get Around and Explore with 5 New Google Maps Updates." *The Keyword (blog)*. *Google*. May 18, 2021. https://blog.google/products/maps/five-maps-updates-io-2021/.

Nakamura, Lisa. "Glitch Racism: Networks as Actors Within Vernacular Internet Theory." *Culture Digitally (blog)*. December 10, 2013. https://culturedigitally.org/2013/12/glitch-racism-networks-as-actors-within-vernacular-internet-theory.

Nicas, Jack. "As Google Maps Renames Neighborhoods, Residents Fume." *The New York Times*, August 2, 2018. https://www.nytimes.com/2018/08/02/technology/google-maps- neighborhood-names.html.

Noble, Safiya. *Algorithms of Oppression: How Search Engines Reinforce Racism*. New York, NY: NYU Press, 2018.

Noble, Safiya U., and Sarah T. Roberts. "Through Google-Colored Glass(es): Design, Emotion, Class, and Wearables as Commodity and Control." In *Emotions, Technology, and Design*, edited by Sharon Y. Tettegah and Safiya Umoja Noble, 187–212. New York, NY: Elsevier, 2016.

Noone, Rebecca. "From Here To: Everyday Wayfinding in the Age of Google Maps." PhD diss., University of Toronto, 2020.

Noone, Rebecca. "Locating Embodied Forms of Urban Wayfinding: An Exploration," *Diversity, Divergence, Dialogue: Proceedings to the iConference 2021*, online, 635–644.

Noone, Rebecca. "Navigating the Threshold of Information Spaces: Drawing and Performance in Action." In *Visual Research Methods: An Introduction for Library and Information Studies*, edited by Shailoo Bedi and Jenaya Webb, 169–188. London: Facet Publishing, 2020.

Noone, Rebecca. *From Here to Drawings*. 2014. https://www.thereroutingproject.org/from-here-to-rebecca-noone.

Noone, Rebecca, and Arun Jacob. "Bad Boundaries: Geofencing and Infrastructures of Trespassing". *TOPIA: The Canadian Journal of Cultural Studies* 48, no. 1, (2024): 76–92.

November, Valerie, Eduardo Camacho-Hübner, and Bruno Latour. "Entering a Risky Territory: Space in the Age of Digital Navigation." *Environment and Planning D: Society and Space* 28, no. 4 (2010): 581–599. 10.1068/d10409.

O'Beirne, Justin. "A Year of Google & Apple Maps: How Much Do Google & Apple Maps Change in a Year?" *Justin O'Beirne*. 2017. https://www.justinobeirne.com/a-year-of-google-maps-and-apple-maps. Accessed May 23, 2021.

O'Beirne, Justin. "Google Maps's Moat: How Far Ahead of Apple Maps is Google Maps?" *Justin O'Beirne*, 2017. https://www.justinobeirne.com/google-maps-moat. Accessed May 23, 2021.

O'Beirne, Justin. "What Happened to Google Maps?" *Justin O'Beirne*. 2016. https://www.justinobeirne.com/what-happened-to-google-maps. Accessed May 23, 2021.

Olson, Parmy. "The 'Father of GPS' Really Doesn't Like Having His Location Tracked." *Forbes*, February 13, 2019. https://www.forbes.com/sites/parmyolson/2019/02/13/the-father-of-gps-really-doesnt-like-having-his-location-tracked/?sh=217fecee5afa.

Oman, Susan. *Understanding Well-Being Data: Improving Social and Cultural Policy, Practice and Research*. London, UK: Palgrave MacMillan, 2021.

Oyedemi, Toks Dele. "Digital Coloniality and 'Next Billion Users': the Political Economy of Google Station in Nigeria." *Information, Communication & Society* 24, no. 3 (August 2020): 329–343. 10.1080/1369118X.2020.1804982

Özkul, Didem. "The Algorithmic Fix: Location Intelligence, Placemaking, and Predictable Futures." *Convergence* 27, no. 3 (April 5, 2021): 594–608. 10.1177/13548565211005644.

Özkul, Didem. "Location as a Sense of Place: Everyday Life, Mobile and Spatial Practices in Urban Spaces." In *Mobility and Locative Media: Mobile Communication in Hybrid Spaces*, edited by Adriana de Souza e Silva and Mimi Sheller, 101–116. New York: Routledge, 2015.

Packer, Jeremy. "What Is an Archive? An Apparatus Model for Communications and Media History." *The Communication Review* 13, no. 1 (2010): 88–104.

Packer, Jeremy. *Mobility Without Mayhem: Safety, Cars, and Citizenship*. Durham, NC: Duke University Press, 2008.

Palaniappan, Meena. "Ditching Diesel: Community-Driven Research Reduces Pollution in West Oakland." *Race, Poverty & the Environment* 11, no. 2 (2004): 31–34. http://www.jstor.org/stable/41554456.

Parker, Martin. "Containerisation: Moving Things and Boxing Ideas." *Mobilities* 8, no. 3 (2012): 1–20.

Parks, Lisa, and Caren Kaplan, eds. *Life in the Age of Drone Warfare*. Durham, NC: Duke University Press, 2017.

Parks, Lisa. "Satellite Views of Srebrenica: Tele-Visuality and the Politics of Witnessing," *Social Identities* 7, no. 4 (2001): 585–611.

Perec, Georges. *An Attempt at Exhausting a Place in Paris* translated by Marc Lowenthal, reprint edition. Wakefield Press, 2010.

Perec, Georges. "Approaches to What?" In *Georges Perec, Species of Spaces and Other Pieces*, translated by John Sturrock, 205–207. London: Penguin Books, 1997.

Perkins, Chris. "Cartography: Progress in Tactile Mapping." *Progress in Human Geography* 26, no. 4 (2002): 521–530. 10.1191/0309132502ph383pr.

Perkins, Chris, and Martin Dodge. "Satellite Imagery and the Spectacle of Secret Spaces." *Geoforum* 40, no. 4 (2009): 546–560.

Petrow, Steven. "I Was a GPS Zombie. Here's What Happened When I Went Back to Paper Maps and Serendipity." *USA Today*, May 22, 2018. https://www.usatoday.com/story/tech/columnist/stevenpetrow/2018/05/22/gps-addict-so-tried-driving-without-apple-google-maps/629814002/.

Pew Research Centre. "The Smartphone Difference." April 1, 2015. https://www.pewresearch.org/internet/2015/04/01/us-smartphone-use-in-2015/

Phillips, Chris. "New Ways Maps Is Getting More Immersive and Sustainable." *The Keyword (blog)*. Google. February 8, 2023. https://www.blog.google/products/maps/sustainable-immersive-maps-announcements/.

Phillips, Chris. "4 New Updates That Make Maps Look and Feel More Like the Real World." *The Keyword (blog)*. Google. September 28, 2022. https://blog.google/products/maps/4-new-updates-maps-searchon-2022/.

Pichai, Sundar. "HBD Maps! Reflecting on 15 Years of Mapping the World," *The Keyword (blog)*. Google. February 6, 2020. https://www.blog.google/perspectives/sundar-pichai/google-maps-15th-birthday/.

Pickles, John. *A History of Spaces: Cartographic Reason, Mapping, and the Geo-Coded World*. London, UK: Routledge, 2004.

Pickles, John. *Ground Truth: The Social Implications of Geographic Information System*. New York: Guilford Press, 1995.

Piepzna-Samarasinha, Leah Lakshmi. *Care Work: Dreaming Disability Justice*. Arsenal Pulp Press, 2021.

Pierce, David. "Google Maps' New 'Immersive View' Combines Street View with Satellites." *The Verge*, May 11, 2022. https://www.theverge.com/2022/5/11/23067016/google-maps-immersive-view-street-satellites.

Plantin, Jean-Christophe. "Google Maps as Cartographic Infrastructure: From Participatory Mapmaking to Database Maintenance," *International Journal of Communication* 12 (June 25, 2018): 489–506.

Plantin, Jean-Christophe, Carl Lagoze, Paul Edwards, and Christian Sandvig. "Infrastructure Studies Meet Platform Studies in the Age of Google and Facebook." *New Media & Society* 20, no. 1 (August 2016): 293–310. 10.1177/1461444816661553.

Pratt, Mary Louise. *Imperial Eyes: Travel Writing and Transculturation*. New York: Routledge, 1992.

Presner, Todd Samuel, David Shepard, and Yoh Kawano. *HyperCities: Thick Mapping in the Digital Humanities*. Cambridge, MA: Harvard University Press, 2014.

Pritchett, Dan. "A Look at How We Tackle Fake and Fraudulent Contributed Content." *The Keyword (blog)*. Google. February 18, 2021. https://blog.google/products/maps/google-maps-101-how-we-tackle-fake-and-fraudulent-contributed-content/.

Radical Access Mapping Project. https://radicalaccessiblecommunities.wordpress.com/.

Rault, Jas. "Window Walls and Other Tricks of Transparency: Digitality, Colonial and Architectural Modernity." *American Quarterly* 72, no. 4 (2021): 937–960.

Reid, Elizabeth. "How 15 Years of Mapping the World Makes Search Better," *The Keyword (blog)*. Google. December 4, 2020. https://blog.google/products/maps/15-years-of-mapping-the-world-makes-search-better.

Reuters. "Cambodia Blasts Google Map of Disputed Thai border." February 5, 2010. https://www.reuters.com/article/idUSSGE61406G20100205.

Reynolds, Kim. "Extraction as White Supremacy: Moving Towards Anti-Extraction Practices in the Arts." *Creative Knowledge Resources*, December 14, 2021. https://www.creativeknow.org/bopawritersforum/extraction-as-white-supremacy.

Risam, Roopika. *New Digital Worlds: Postcolonial Digital Humanities in Theory, Praxis, and Pedagogy*. Chicago: Northwestern University Press, 2018.

Roberts, Sarah T. *Behind the Screen: Content Moderation in the Shadows of Social Media*. New Haven, CT: Yale University Press, 2019.

Roberts, Sarah T. "Social Media's Silent Filter." *The Atlantic*, March 8, 2017. https://www.theatlantic.com/technology/archive/2017/03/commercial-content-moderation/518796/.

Roberts, Susan M. and Richard H. Schein. "Earth Shattering: Global Imagery and GIS". In *Ground Truth: The Social Implications of Geographic Information Systems*, edited by Pickles John, 171–195. New York, NY: Guildford, 1995.

Robinson, Arthur H., Joel L. Morrison, Phillip C. Muehrcke, A. Jon Kimerling, and Stephen C. Guptill. *Elements of Cartography*, 6th Edition. Hoboken, NJ: Wiley, 1995.

Robinson, Cedric. *Black Marxism: The Making of the Black Radical Tradition*, 2nd Edition. London: Zed Books, 1983.

Romm, Cari. "Using Google Maps Too Much Really Does Mess with Your Sense of Direction." *The Cut*, January 10, 2017. https://www.thecut.com/2017/01/using-gps-really-does-mess-with-your-sense-of-direction.html.

Rüegg, Arthur (ed.). *René Burri. Brasilia: Photographs 1958–1997*. Zurich: Scheidegger & Spies, 2011.

Russell, Ethan. "How Google Street View Mapped the World." Produced by Wired. June 20, 2022. YouTube video, 9:38. https://www.youtube.com/watch?v=oApM0jBRKbY

Russell, Legacy. *Glitch Feminism: A Manifesto*. Cambridge, MA: Polity, 2020.

Russeth, Andrew. "Stanley Brouwn, Whose Worse Examine Measurement and Memory, Dies at 81," *Art News*, May 22, 2017.

Safransky, Sara. "Geographies of Algorithmic Violence: Redlining the Smart City." *International Journal of Urban and Regional Research* 44, no. 2 (November 2019): 200–218. 10.1111/1468-2427.12833.

Sattiraju, Nikitha. "The Secret Cost of Google's Data Centers: Billions of Gallons of Water to Cool Servers." *Time*, April 2, 2020. https://time.com/5814276/google-data-centers-water/.

Sevigny, Melissa. "Navajo Nation Homes Get Addresses from Google Mapping Project." *KNAU News Talk, Arizona Public Radio*. January 24, 2020. https://www.knau.org/knau-and-arizona-news/2020-01-24/navajo-nation-homes-get-addresses-from-google-mapping-project?fbclid=IwAR1j5MsjzXH6H3-fdQs5FBgMhRz4SLb-3IVV2IR4NgprMY-qzKAPKI2y1qQU.

Shapiro, Aaron. "Street-Level: Google Street View's Abstraction by Datafication." *New Media & Society* 20, no. 3 (January 2017): 1201–1219. 10.1177/146144481 6687293.

Sharma, Sarah, and Armond R. Towns. "Ceasing Fire and Seizing Time: LA Gang Tours and the White Control of Mobility." *Transfers: Interdisciplinary Journal of Mobility Studies* 6, no. 1 (2016): 26–44.

Sharma, Sarah. *In the Meantime: Temporality and Cultural Politics*. Durham, NC: Duke University Press, 2014.

Sharma, Sarah. "It Changes Space and Time! Introducing Power-Chronography." In *Communication Matters: Materialist Approaches to Media, Mobility and Networks*, edited by Jeremy Packer and Stephen B. Crofts, 66–77. New York: Routledge, 2012.

Shekhar, Shashi, and Pamela Vold. "Geographic Information Systems and Cartography." In *Spatial Computing*, 91–125. Cambridge, MA: MIT Press, 2020.

Sieber, Renée and Mordechai Haklay. "The Epistemology(s) of Volunteered Geographic Information: A Critique." *GEO: Geography and Environment* 2, no. 2 (2015): 122–136. 10.1002/geo2.10;

Siegert, Bernhard. "The Map Is the Territory." *Radical Philosophy* 169, 2011 https://www.radicalphilosophy.com/article/the-map-is-the-territory.

Simpson, Audra. "On Ethnographic Refusal: Indigeneity, 'Voice' and Colonial Citizenship." *Junctures: The Journal for Thematic Dialogue* 9 (2007): 67–80.

Simpson, Leanne Betasamosake. "Indigenous Resurgence and Co-Resistance." *Critical Ethnic Studies* 2, no. 2 (2016): 19–34. 10.5749/jcritethnstud.2.2.0019.

Simpson, Leanne Betasamosake. *As We Have Always Done: Indigenous Freedom Through Radical Resistance*. Minneapolis: University of Minnesota Press, 2017.

Singh, Rianka. "Resistance in a Minor Key: Care, Survival and Convening on the Margins." *First Monday* 25, no. 5 (May 2020). 10.5210/fm.v25i5.10631

Singh, Rianka and Sarah Banet-Weiser. "Sky High: Platforms and the Feminist Politics of Visibility." *Re-Understanding Media: Feminist Extensions of Marshall McLuhan*, ed. Sarah Sharma and Rianka Singh. Durham, NC: Duke University Press, 2022, 163–175.

Smith, Andrew. "Leveraging Sport Mega-Events: New Model or Convenient Justification?" *Journal of Policy Research in Tourism, Leisure and Events* 6, no. 1 (August 1, 2013): 15–30. 10.1080/19407963.2013.823976.

Smith, Linda Tuhiwai. *Decolonizing Methodologies: Research and Indigenous Peoples*, 2nd Ed. London, UK: Zed Books Ltd., 2013.

Sofia, Zoë. "Container Technologies." *Hypatia* 15, no. 2 (March 2020): 181–201. 10.1111/j.1527-2001.2000.tb00322.x.

Spade, Dean. "Solidarity Not Charity: Mutual Aid for Mobilization and Survival." *Social Text* 38, no. 1 (2020): 131–151.

Spade, Dean. *Normal Life: Administrative Violence, Critical Trans Politics, and the Limits of Law*. 2nd Edition. Duke University Press, 2015.

Spade, Dean. *Mutual Aid: Building Solidarity During This Crisis (and the Next)*. London, UK: Verso, 2020.

Squires, Gregory D. "Racial Profiling, Insurance Style: Insurance Redlining and the Uneven Development of Metropolitan Areas." *Journal of Urban Affairs* 25, no. 4 (2003): 391–410. 10.1111/1467-9906.t01-1-00168.

Strauss, Claudia. "The Imaginary." *Anthropological Theory* 6, no. 3 (September 2006): 322–344. 10.1177/1463499606066891.

Suarez, Joanna. "Google Maps Mistakes N.J. House for State Park." *ABC News*, July 12, 2011. https://abcnews.go.com/Technology/google-maps-mistakes-jersey-house-state-park/story?id=14056722.

Suchman, Lucy. "Anthropological Relocations and the Limits of Design." *Annual Review of Anthropology* 40, no. 1 (2011): 1–18.

144   *Bibliography*

Sutherland, Tonia. Making a Killing: On Race, Ritual, and (Re)Membering in Digital Culture. *Preservation, Digital Technology & Culture* 36, no. 1 (2017): 32–40. 10.1515/pdtc-2017-0025

Sutter, John D. "Google Maps Border Becomes Part of International Dispute." *CNN News*, November 5, 2010. http://www.cnn.com/2010/TECH/web/11/05 /nicaragua. raid.google.maps/index.html.

Sweeney, Don. "Google Maps Sends Hikers on 'Potentially Fatal' Route in Scotland, Climbers Say." *The Charlotte Observer*, July 18, 2021. https://www.msn.com/en-us/ travel/tripideas/google-maps-sends-hikers-on-e2-80-98potentially-fatal-e2-80-99-route-in-scotland-climbers-say/ar-AAMiv6q.

Talbot, Adam. "Vila Autódromo: The Favela Fighting Back Against Rio's Olympic Development." *The Conversation*, January 12, 2016. https://theconversation.com/ vila-autodromo-the-favela-fighting-back-against-rios-olympic-development-52393.

Tarkka, Minna. "Labours of Location: Acting in the Pervasive Media Space." In *The Wireless Spectrum: The Politics, Practices, and Poetics of Mobile Media*, edited by Barbara Crow, Michael Longford, and Kim Sawchuk, 131–145. Toronto: University of Toronto Press, 2010.

Terez, Kelly, and Brad Mielke. "Google Maps Shortcuts in Colorado Turns into a 'Muddy Mess' with a 'Hundred Cars.'" *ABC News*, June 26, 2019. https://abcnews. go.com/US/google-maps-shortcut-colorado-turns-muddy-mess-hundred/story?id= 63946068&cid=clicksource_4380645_null_twopack_hed.

Thatcher, Jim, Luke Bergmann, Britta Ricker, Reuben Rose-Redwood, David O'Sullivan, Trevor J. Barnes, Luke R. Barnesmoore, et al. "Revisiting Critical GIS." *Environment and Planning A* 48, no. 5 (2015): 815–824.

Thiagarajan, Shiva, and Rio Akasaka. "Building a Map for Everyone." *The Keyword (blog)*. Google. July 6, 2017. https://blog.google/products/maps/building-map-everyone/.

Thiel, Tamiko. *Clouding Green*. Augmented reality installation. Palo Alto, California, 2012. https://tamikothiel.com/projects-vr-ar-installations.html.

Tidy, Joanna. "The Gender Politics of 'Ground Truth' in the Military Dissent Movement: The Power and Limits of Authenticity Claims Regarding War." *International Political Sociology* 10, no. 2 (June 2016): 99–114. https://doi.org/10. 1093/ips/olw003.

Tonkiss, Fran. *Space, the City, and Social Theory*. Cambridge, UK: Polity, 2004.

Transport for London. "Plan a Journey." https://tfl.gov.uk/plan-a-journey/. Accessed June 23, 2023.

Tuck, Eve, and K. Wayne Yang. "Decolonization Is Not a Metaphor." *Decolonization: Indigeneity, Education & Society* 1, no. 1 (2021): 1–40.

Tuck, Eve, and K. Wayne Yang. "Unbecoming Claims: Pedagogies of Refusal in Qualitative Research." *Qualitative Inquiry* 20, no. 6 (2014): 811–818.

Tuck, Eve, and Marcia McKenzie. *Place in Research: Theory, Methodology, and Methods*. London, UK: Routledge, 2015. Proquest. http://site.ebrary.com/id/ 10941500.

University of California, Berkley. "Most to Least Segregated Cities in 2020: According to 2020 Census Data." Othering and Belonging Institute. 2020. https:// belonging.berkeley.edu/most-least-segregated-cities-in-2020.

Vaidhyanathan, Siva. *The Googlization of Everything (and Why We Should Worry)*. Los Angeles, CA: University of California Press, 2011.

van der Meijden, Peter A. *This Way Bruown: The Archive—Past, Present, and Future*. Performing Archives/Archives of Performance. Edited by Gunhild Borggreen and Rune Gade. Copenhagen, DK: Museum Tusculanum Press, 2013.

van Doorn, Niels. "Platform Labor: On the Gendered and Racialized Exploitation of Low-Income Service Work in the 'On-Demand' Economy." *Information,*

*Communication & Society* 20, no. 6 (2017): 898–914. 10.1080/1369118X.2017. 1294194.

Verhoeff, Nanna. *Mobile Screens: The Visual Regime of Navigation.* Amsterdam: Amsterdam University Press, 2012.

Vertesi, Janet. "Mind the Gap: The London Underground Map and Users' Representations of Urban Space." *Social Studies of Science London* 38, no. 1 (2008): 7–33.

Virilio, Paul. *War and Cinema: The Logics of Perception.* London, UK: Verso, 1989.

Waidner, Isabel. *Sterling Karat Gold.* London: Peninsula Press, 2021.

Walcott, Rinaldo. *On Property: Policing, Prisons, and the Call for Abolition.* Windsor, ON: Biblioasis, 2021.

Walia, Harsha. *Border and Rule: Global Migration, Capitalism, and the Rise of Racist Nationalism.* Halifax, NS: Fernwood Publishing, 2021.

Weckert, Simon. "Google Maps Hack" Performance and Installation (2020). https://simonweckert.com/googlemapshacks.html

West, Sarah Myers. "Censored, Suspended, Shadowbanned: User Interpretations of Content Moderation on Social Media Platforms." *New Media & Society* 20, no. 11 (2018). 10.1177/1461444818773059.

Whipple, Tom. "GPS Creator Bradford Parkinson Mourns Lost Art of Map Reading." *The Times*, February 13, 2019. https://www.thetimes.co.uk/article/gps-creator-bradford-parkinson-mourns-lost-art-of-map-reading-2xsvkb8x6.

Wilken, Rowan, and Julian Thomas. "Vertical Geomediation: The Automation and Platformization of Photogrammetry." *New Media & Society* 24, no. 11 (2022): 2531–2547. 10.1177/14614448221122214.

Wilken, Rowan. *In Cultural Economies of Locative Media.* London: Oxford University Press, 2019.

Wilken, Rowan, and Gerard Goggin, eds. *Locative Media*—Definitions, Histories, Theories. In *Locative Media*, edited by Rowan Wilken and Gerard Goggin, 1–19. New York: Routledge, 2014.

Wong, Stephanie. "A Sustainable Solution Helped a Small Town Cool Its Data Center." *The Keyword (blog)*. Google. May 31, 2023. https://blog.google/inside-google/infrastructure/a-sustainable-solution-helped-a-small-town-cool-its-data-center/.

Young, Liam Cole. "Cultural Techniques and Logistical Media: Tuning German and Anglo-American Media Studies." *M/C Journal* 18, no. 2 (April 29, 2015). https://www.journal.media-culture.org.au/index.php/mcjournal/article/view/961.

Zuboff, Shoshana. *The Age of Surveillance Capitalism: The Fight for a Human Future at the New Frontier of Power*, 1st trade paperback edition. New York, NY: PublicAffairs, 2020.

Zukin, Sharon. *Naked City: The Death and Life of Authentic Urban Places.* Oxford University Press, 2010.

Zukin, Sharon, Scarlett Lindeman, and Laurie Hurson. "The Omnivore's Neighborhood? Online Restaurant Reviews, Race, and Gentrification." *Journal of Consumer Culture*, 17, no. 3 (2015): 459–479. 10.1177/1469540515611203.

# Index

Pages in *italics* refer to figures and pages followed by n refer to notes.

For Product Safety Concerns and Information please contact our EU
representative GPSR@taylorandfrancis.com
Taylor & Francis Verlag GmbH, Kaufingerstraße 24, 80331 München, Germany